ON BEING HUMAN

On Being Human constitutes an achievement in anthropological analysis, conceived as a broad interdisciplinary endeavour, and encompassing a huge range of theoretical and practical issues. Marsh offers a multi-level exposition on the scope of human nature, added to by scientifically and bio-medically based approaches, thus permitting an accurate assessment of personhood and its defining characteristics. This is the necessary basis for a serious and engaged treatment of very sensitive issues concerning human life, its beginnings, its end, and dignity (particularly for the fetus, the disabled, and infirm). Marsh brings the ethical discussion to a new high with its impressive and wide erudition, employing well-reasoned arguments from recent research, to assist in the best understanding of the meaning of human life – and how to live it.

Fr. Lluis Oviedo OFM, Professor of Christian Anthropology, Pontifical University Antonianum, Rome, Italy

After a long and distinguished medical career, where one's identity and sense of self develops primarily through what one does, Michael Marsh takes the reader on a scientifically informed, spiritual journey of Christian faith in search of what it means to simply be. An inspiring and sometimes challenging, personal interpretation of the sanctity and meaning of being human. Highly recommended.

Jay R. Feierman MD, Professor of Psychiatry (Emeritus), University of New Mexico, USA

Having spent a lifetime in clinical practice and biomedical research, Michael Marsh argues that routine discussions of what it is to be human too rarely consider the large proportion of humankind who are disfigured or disabled, dysfunctional or approaching death. In a deeply empathetic discussion, and writing as both theologian and scientist, he provides an analysis that successfully avoids the hyperinflation of the nobler aspects of creation and of humankind. He is critical of conversation-stopping appeals to our being made in the image of God and to secular moral philosophies that render the severely incapacitated simply disposable. I warmly recommend his book for the wisdom it enshrines on such controversial issues as abortion, infanticide and assisted dying. It will have a special appeal for Christian moralists because it points to the possibility of humans becoming "divinised in the Godhead" beyond this world.

John Hedley Brooke, The Andreos Idreos Professor (Emeritus) of the History of Science, and Fellow, Harris Manchester College, University of Oxford, UK

On Being Human

Distinctiveness, Dignity, Disability & Disposal

On Being Human

Distinctiveness, Dignity, Disability & Disposal

Michael N. Marsh

BOOKS

Winchester, UK
Washington, USA

First published by iff Books, 2015
iff Books is an imprint of John Hunt Publishing Ltd., Laurel House, Station Approach,
Alresford, Hants, SO24 9JH, UK
office1@jhpbooks.net
www.johnhuntpublishing.com
www.iff-books.com

For distributor details and how to order please visit the 'Ordering' section on our website.

Text copyright: Michael N. Marsh 2014

ISBN: 978 1 78279 451 6
Library of Congress Control Number: 2015932905

A CIP catalogue record for this book is available from the British Library.

Design: Lee Nash

Printed and bound by CPI Group (UK) Ltd, Croydon, CR0 4YY, UK

We operate a distinctive and ethical publishing philosophy in all
areas of our business, from our global network of authors to
production and worldwide distribution.

CONTENTS

Duncan,

Thank you for your
comments on some of
the material included
within.

With all best wishes

Michael.

19. IX. 2015.

Preface

This book originates from a series of public lectures given in Hilary Term, 2010 under the auspices of The Oxford Centre for Christianity and Culture (OCCC), Regent's Park College, University of Oxford. They were given under the general title 'Human Uniqueness and Dignity', based on the presumption of man-made-in-God's image as proclaimed in the first chapters of Genesis. Yet I deliberately turned that precept round to ask 'Are we made in God's image?' and if so, what makes us so special in enabling us to claim God-like mirroring for ourselves. This would be in direct opposition to the rather less impressive and possibly more mundane descent, albeit direct (pace Thomas Henry Huxley), from apes and monkeys.

Many books which have dealt with this subject in the past, often written by theologians or philosophers, have tended to honour creation and humankind, thus to envision the latter as standing in an elevated position. My own approach as the title suggests, based largely on a lifetime in clinical practice and biomedical research, sees things a little differently. Of course there is a grandeur both in all that we see around us and in how the human body functions. But life is also hard and very gritty, a feature which, although it affects large numbers of individuals, somehow never directly enters the minds of those for whom the ravages of disease or environmental disaster seem not to be part of their daily routine, commerce, or thought.

Thus my approach draws on four themes: *Distinctiveness* arising from evolutionary anthropology, coupled with further observations on genes, consciousness and language. That is compared and expanded with a theological anthropology of humankind, which I term *Dignity*. Having drawn those distinctions, I then exemplify those features of life which reflect its less attractive aspects. *Disability* refers to the factors intrinsic and extrinsic to varied states of discomfort which are not always bracketed within this designation, while *Disposal* deals with the ways in which death for those at the extremes of life are viewed and treated by society at large, and for which the current problem of assisted suicide looms menacingly large.

This book does not have to be read from cover to cover. Each essay stands on its own. There is, however, a sense of development in passing through aspects of evolutionary history, genes, language and consciousness to a theological sketch of humankind. Because each chapter is a self-sufficient unit, there has been some reduplication of material, so as obviously not to have the reader continually moving from one chapter to the other in order to make sense of a theme. That sentiment applies to aspects of embryology seen in the chapters on the dignity of the human embryo, and again when I deal with abortion and infanticide. Also in the chapters dealing with the metaphysical aspects of what a human being, ultimately, is.

I am most grateful to Dr. Nicholas Wood, Dean of Studies and Director of OCCC, Regent's Park College for allowing me this series of lectures, advertising them in the Oxford Diocesan newspaper *The Door*, and for making the appropriate collegiate arrangements for their delivery.

There are many university colleagues who have made comments, given advice or helped either in contributing to the book, or more importantly, clarifying my own thoughts – here and there. But special grateful thanks go to my friends and colleagues Drs Peter Colyer and Duncan Forbes, whose steadying hands provided well-received assistance during the course of my assembling of the final manuscript.

Michael N. Marsh, Wolfson College, and The Oxford Centre for Christianity & Culture, Regent's Park College, University of Oxford.

Part I

Distinctiveness

Part 3

Distinctiveness

Chapter 1

Introduction: Exploring the Breadth & Depth of Being Human

This book derives from public lectures given in Oxford during Hilary Term 2010, based on the presumption of man-in-God's image as proclaimed in the first chapters of Genesis. Yet I deliberately turned that precept round to enquire – 'Are we made in God's image?' – and if so, what characteristic(s) make us so special in claiming that divine resemblance. The alternative, perhaps a little less impressive, relies on origins from apes. This book elucidates our evolutionary origins and status as human beings: but, in addition, it enlarges the curriculum in dealing not only with the various disabilities which confront us in this world, but also with our beginnings and endings, including abortion and infanticide, as well as the legalities and theology of assisted dying. The chapters included thereby provide a comprehensive picture of human life as lived by many on this planet.

The reader is first invited to become conversant with human physical anthropology through evolutionary time (Chapter 2), and the biological specifics contributing to each individual life-history, through genes (Chapter 3), conscious-awareness (Chapter 4) and the acquisition of language (Chapter 5). Those aspects define Distinctiveness. In dealing with our physical evolutionary origins, I suggest that we are more distinctive than 'unique', relative to the higher apes. Despite sharing surprisingly closely linked genetic endowments (genomes) with them, our bodily features (phenotypes) are markedly different, as well as our cultural differences.

The widely held belief in human uniqueness is perhaps a biblical hangover from us being supposedly made in the image of God. The problem is in knowing what that axiom precisely

3

conveys. Evolutionary novelty usually emerges from re-use of preceding material. Thus, in the step from invertebrate to vertebrate the existing genome was doubled thereby providing scope for newer, original developments. Humans do not possess genes that have arisen *de novo*, instead through genetic mutations and reduplications, helped by sexual forms of reproduction which reconfigure the genetic elements or 'genotype' contributed, and which influence bodily configuration or phenotype. Nowadays, we should realise that defining a 'gene' is very difficult (Chapter 3: Genes). Genes often work in 'groups', with multi-genes controlling the activity of many others in a different part of the DNA sequence.

Environmental influences on gene expression are being seen to play increasingly more important roles in their overall (Darwinian) activity and control. Thus the field of 'epigenetics' (non-Darwinian inheritance) relating to environmental effects is now emerging as an overwhelming super-dominant element in the way through which genes are activated and regulated, and thus how they are expressed. We come upon this in Chapter 7 (Can We Ascribe Moral Status to the Human Embryo-Foetus?), for example, where epigenetic influences witnessed through The Dutch Winter of Starvation, November 1944-April 1945; the impressive Avon County (Bristol, UK) Long-Term study of parents and children (ALSPAC); and the Swedish long-term observations of families in terms of dietary effects on the future lives of grandchildren and their health, have been documented with some surprising outcomes. Some of these latter effects, amazingly, are manifested through the male (paternal) line, bringing further important *moral* issues to play in relation to a prospective embryo-foetus *in utero*, apart from the more obviously well-known strictures applicable to women during the period of gestation itself.

We are beginning to progress beyond the idea that humans are driven by 'selfish' genes, and, derivative of that outmoded idea, the similarly overblown concepts of Evolutionary Biology, Evolutionary-Sociology, Evolutionary-Psychology and other neo-Darwinistic 'Evo-this-and-that' things. Collectively, Raymond Tallis, and a former medical colleague of mine, refer to much of that as 'evo-rubbish'.[1] The philosopher Roger Scruton dismisses it as 'neurononsense'.[2] Nevertheless, the increasing role of environment in configuring what we become is of far-reaching importance to our understandings of bodily function and hence of personhood.

We do not have consciousness, we *are* conscious: that is, each individual is a conscious being, being conscious (Chapter 4: On Being Conscious). I am unsure whether science can solve the emergence of this quality from an astoundingly complex brain possessing around 30 billion nerve cells (known as neurones) and an even astronomically higher number of inter-neuronal connections, or 'synapses'. We know that a single neurone may be attached synaptically to as many as 10,000 other neurones, resulting in phenomenally intricate connections throughout the brain. Indeed, it is the vast and unimaginable synaptic connectivity in our brains which confers humankind's outstanding levels of cognitive and affective capacity.

However, the major conundrum centres on how subjective conscious-awareness is able to control an objective physical organ (when thinking, for example) when consciousness itself originates from that physical organ. Clearly there is a very tight connection between neural processes and cognitive outcomes about which we can only have speculative ideas. We need to be aware that the brain operates at several different levels, concurrently, but not necessarily within our conscious-awareness. Next, I touch on the evolutionary history of consciousness, and further consider when conscious-awareness develops in the foetal brain – an aspect of consciousness hardly considered in many other books on brain function, but functionally important. Despite the restricted lack of sensory input which obtains in the womb, it is evident that much conscious activity is present, at least during the last ten weeks (probably longer) of gestation: there is no such thing as a 'blank slate' at birth.

Apart from that, we all have the innate ability to render ourselves unconscious when we sleep: likewise when undergoing general anaesthesia. Many new insights about being conscious have arisen from the various studies into these latter phenomena.

On the other hand, much of learned conscious activity operates efficiently at sub-conscious level: so, why do we need to be conscious? For example, just consider the enormous success through evolutionary time demonstrated by bees, wasps, spiders and ants! Secondly, what is the evolutionary advantage of consciousness requiring brains to absorb 25% daily energy intake for their operation? That was an enormous biological gamble even albeit blindly accomplished – if we are true Darwinists. So how could that gamble have been prefigured in the progressive, blind mechanisms of selection? Thirdly, from some quarters, neurophysiological doubts about 'freewill' and our supposed ability to operate independently with our own choices have arisen. This leads on to considerations as to how we derive a sense of (and even believe in) the divine. Is the latter pure artefact conjured up by a brain that controls mind? Reflection on these many issues has an important impact on our belief systems and whether concepts of the divine, in the end, turn out to be mere cerebral make-believe.

Despite our sharing ~95% genetic homology with apes, why do humans speak? It is an extraordinarily novel development due, in part, to two key mutations involving the FOXP2 gene which arose during the relatively brief evolutionary period following our separation from chimpanzees, ~5 million years ago. That compares with the remarkably highly conserved (non-mutational) behaviour of this gene as far back as the origins of mice, ~100 million years ago – and even perhaps further back throughout evolutionary time (Chapter 5: On Having Language). It is highly pertinent to ask how and why a single amino acid substitution (from one of two mutations) should have come about and resulted in the foundations of articulatory speech and how that chance evolutionary change, an almost miniscule alteration, now underpins and has therefore shaped our highest intellectual faculties.

Speech is an indispensable requirement for abstract thinking, the use of metaphorical language, and the basis for religious beliefs. Despite occasional excited newspaper reports of so-called 'God-spots' in the brain, I pointed out some years back that there is no locus for a holy shrine within the brain, sheltered from the common vulgarities of secularly-based neural commerce.[3] Our acceptance of divine presence and action is predominantly cognitively based on a faith in that divinity, and buttressed by forward-looking hope and belief in the propositions thereby arising. Apart from that there is, additionally, an inexorable evolutionary movement towards higher biological complexity, size, and burgeoning functional outcomes. This can be seen as a type of 'progression' (not of Enlightenment kind) towards the ultimate emergence of human beings, their big brains, and their extraordinary endowments, capacities and capabilities.

Next, I turn to Dignity, and its relevance to theological concepts of being human, and in the making of personhood. While genes, consciousness, and language certainly frame mankind within the evolutionary environment as well as the physical world which mankind inhabits, more than a purely physical anthropology is needed. In part, that includes the insatiable thirst for transcendence by bursting through barriers to conquer what lies on the other side. But in that pursuit of transcendence, we step outside ourselves (ekstasis): and through this action we discover the divine (Chapter 6: Theological Anthropology). The term Dignity underpins theological aspects of humankind, seen as a means of defining the true nature of humankind as persons, but of persons in relation with each other and the Godhead.

As this chapter will explain, our createdness imposes certain restrictions, in that we cannot escape the boundaries of our world – nor fully understand ourselves or others. That dilemma creates the further paradox of 'presence-in-absence'. And since all other humans are made in the same mould, they cannot provide answers to, or escape, these frustrating outcomes. The ultimate method of transcendence (the attempt to break out from ourselves and our physical circumstances) is to be in relation with an Other, that is, the Godhead, achieved principally through baptism.

In the early church, baptism was viewed eschatologically (through the Spirit towards the final consummation of creation), making the convert a member of the sacral community (on earth as in heaven) and a part of the Godhead – through 'dying' and 'rising with' Christ. This partial escape from the bondage of creat-edness is subsumed under the notion, within the Orthodox tradition, of 'godliness', or 'theosis' (Chapter 6: Theological Anthropology). This has no relationship to a 'soul', but does constitute a sacralised form of being or existence, apart from the being based on a fleshly body. Baptism thus provides continuity towards, and within, the Godhead once the corporeal self has dwindled and died. The role of baptism as a means of metaphysical re-birth 'from above' in establishing the person as 'unique' and as one known individually by the Godhead, figures heavily in my deliberations.[4] Our guarantee is provided by the Chalcedonian Definition that there can be two (ontological) natures within one (human) person 'without confusion'.[5] And that really is a gift given by the Godhead: we can achieve very little on our own.

The idea of being made in the divine image means becoming God-like, rather than being seen in some other guise or construction. The oft-used expression 'sanctity of the body or flesh', similar to 'made in God's image' are categories which, in somehow being thought to carry with them a kind of divine imprimatur, become barriers (conversation stoppers) which cannot, and should not, be transgressed. In disclosing a sense of divine authority, they tend to inhibit further thoughtful analysis of their meanings, and in my view fail to frame the uniqueness of the person. Therefore, I am unsympathetic to those precepts, which are easily overridden by other more appropriate and sustainable arguments.

For example, many would regard the medical advances through which the pre-existing, diseased body can be made more wholesome, to be an entirely acceptable mode of operation. That has been achieved through surgery, immunotherapy, pharmaceutical procedures, materials science, technological advances and so on, as major components of the clinical armamentarium. Despite their incursions into body and flesh (whether or not considered 'sanctified'), they should be welcomed. Even the Godhead might be pleased with the ingenuity and subtle research enterprises which underpin them. There are few, indeed, who would not have been beneficiaries of such protocols and their skilled application.

In proceeding and reflecting more specifically on the question of defining persons and personhood, we should note how definitions have changed through time, particularly since the Enlightenment.[6] However, those are man-made definitions, which may therefore fail to do justice to defining an individual person. Moreover, the kind of definitions in force at any time considerably influences how we view each other, and how society interprets those definitions and accordingly manipulates its own agendas, jurisprudence and socio-political approaches towards society and individuals.

Current trends have reduced the definitional idea of a person to a collection of abstract qualities, for example, memory, ongoing interests, and future intentions. These definitions float tenuously in space, lacking a suitable ontological basis whether conceived anthropologically, sociologically or theologically. Since persons constantly change over the period of their lifetimes, it is difficult to know when some or all of these properties could be operative, and, of course, when they disappear. On those problematic grounds, a person would only truly come into view once certain arbitrary, pre-determined properties are satisfied. Is that really possible? On the other hand, such criteria exclude many who should be regarded as part of our community – the embryo-foetus, neonates, children, many disabled folk, the mentally ill, and the elderly and frail including those in end-of-life situations – and so are easily dismissed as so-called 'non-persons'. From that position, it is an easy step for those categories of humanity to be regarded as worthless, unworthy and therefore available for disposal. My approach counters that tendency.

This difficulty first impinges on the status of the embryo-foetus (Chapter 7: Can We Ascribe Moral Status to the Human Embryo-Foetus?) and whether it can be regarded as something that is of humankind and valuable. In two notably germane publications on this subject, Norman Ford's *When Did I Begin?* and the UK government-sponsored *Report into Human Fertilisation and Embryology* chaired by Baroness Mary Warnock, it was insisted that a 'person' (undefined by both authors) could only come into existence, and hence be recognised as such, once the last opportunity for twinning had passed and when the primitive streak had become visible on the developing early embryo, both occurring around 2 weeks after fertilisation.[7,8] Here, personhood comes to be defined in terms of two biological phenomena, and nothing more. I am puzzled by how a biological structure can underpin moral decisions, and even the basis of law (Warnock).

In those two publications, all preceding stages in development were dismissed and deemed inconsequential. Some philosophers employ analogous arguments. Employing molecular genetics and biological data, I demonstrate the improbable basis for those conjectures, concluding by appropriate arguments from molecular biology that the new individual with a unique genome starts when most people imagine it to come into being – once fertilisation has taken place.

This is a more important conclusion than it appears at first glance. That is because the early embryo-foetus is not some kind of abstract entity, but a new individual with its own specific genome. That is the origin from which we all derive: it is not a mere inconsequential blob of cells as some apparently think. It is a reflection of the continuity of life from its initial emergence on the planet, since any foetus is *not* a new life, but continues the living-principle (of a living sperm and ovum) into a new, genetically unique, individual. Second, that new individual is conceived within a family setting, is influenced by the lifestyles and defined contingencies affecting its grandparents. Various long-term studies, including observations of current adults (but who were conceived during the intense starvation caused by the German occupation of Holland) demonstrate how the seeds of later adult illness are sown during embryogenesis – and before. These observations are of grave moral concern and should influence, for example, public health programmes. These are the considerations, among others, which impart to that conceptus moral dignity and therefore assign to it a respect which can be easily lost, forgotten or overlooked.

That foetuses *in utero* are now surgically operated upon renders those involved (doctors and allied professionals) susceptible to criteria of care as laid down for any person undergoing a medical procedure, and to the legal outcomes of negligence should there be a fatality. These necessities likewise reveal the foetus in a different light, and in respect of moral accountability.

Conversely, it has been queried: first, how the embryo-foetus (from the earliest two-celled zygote) could sensibly be regarded as a 'person'; second, in the absurdness of supposing that we should be able to extend our psychological continuity back to the earliest stages of the embryo – and if not, then it could not be a 'person' in continuity with my current 'self'; and third, whether an embryo-foetus could realistically be deemed a 'potential person'. Is it not the case that life, throughout its course, invariably offers the precarious balance between potentiality and possibility? That questions of this type in respect of the embryo-foetus can be raised rests entirely on current definitions of what is meant by a person and personhood, and the criteria employed.

Next, the idea of personhood impinges on *Disability*. This focuses not so much on ideas of helplessness as revealed through the foetus, but on disaster, disease and other catastrophes of living life on earth, which afflict people of all ages. This is discussed in Chapter 8 (On Being Disabled, Dysfunctional, and Disfigured). I have already asserted that life is not solely one of grandeur since, seen particularly from a clinical perspective, it is hard, gritty and very debilitating for those caught up within the natural history of prolonged illnesses, whether due to degenerative conditions of the muscles or nervous system, malignancy, metabolic upsets, and/or genetic mutations which lead to multi-system dysfunctions. In highlighting disablement, I bring to the fore a subject which hardly enters into considerations when humanity or personhood, in general, are being spoken about. Yet, this 'underbelly' of humanity involves the majority of those on our planet, in some disabling form or another.

Yet, there is a noticeable divide within society between those who are envisioned as being disabled or dysfunctional, and the remaining 'able-bodied' community. But even thinking that we are able-bodied is erroneous because we are all moving along a slowly degenerative trajectory even though its effects may not immediately be present. In other cases, that dysfunction may be present, although not publicly visible and therefore kept latent by those enduring such effects. The imagined divide, between 'them' and 'us', is grossly mistaken: disability within society does not follow a bimodal distribution.[9]

Disability can take many hues. At the time of writing, it was announced (BBC Television 24-hour News bulletin) that almost 30,000 UK children were admitted to hospital during 2014 as a result of self-harm, while over one third of a million were abused (sexually, violently, verbally, or through personal theft).[10] These traumatic events are occasioned by bullying, adolescent peer pressures, electronic social media platforms, family rifts, and many other factors current in modern society which considerably dis-able the lives of their victims. On a completely different, yet relevant front, the combined threats from Ebola and malarial infection in Africa have claimed the lives of up to 15,000 people during the last nine months (2014-15).[11] Clearly, we need to be made aware of these problems as part of being human. These disasters need to be recognised since apart from the disablement as such, they are a direct threat to existence, especially to dependants in very poor countries when the family breadwinner succumbs.

We inhabit an era of great uncertainty and of conflicting antitheses so that life is not entirely consistent with the rosy picture espoused by Genesis, philosophers or soothsayers. It is easy, I suggest, to hyper-inflate the nobler sides of creation and of mankind. We praise the beauty of creation in one breath, while denouncing natural geophysical disasters with another, without realising that our planet necessarily must produce these often life-shattering effects in the course of its continued working as a natural system. Those disasters hurt, maim, kill, or wreck the lives of millions upon millions of people, and so we need to ask what kind of restoration could ever ensue to neutralise that loss of life or hope of survival.

Finally, Chapters 9, 10 and 11 are concerned with Disposal. These chapters include considerations about abortion and infanticide (Chapter 9) as well as looking at the other extreme of life, where we are concerned with the removal of the infirm or those at the end stage of a life-destroying disease. These actions depend on legal provisions (Chapter 10: Assisted Death: Legal Aspects) and the question whether a theological ground for assisted deaths can be evolved (Chapter 11: Assisted Death: Theological Proposals).

Resulting from current trends throughout society, it comes as no surprise that embryos are destroyed at will, foetuses aborted, children murdered, the mentally ill left untreated, and the elderly and infirm abused at home, in care facilities or the geriatric wards of our hospitals. Proof of these abuses can be read about daily in our newspapers. More importantly, a dangerous trend has arisen leading to attitudes of moral irrelevance regarding those not included in the listings of the very restricted, abstract, but desirable features attributed to 'proper persons' and their so-called 'moral community' as defined by trendy philosophers. This creates a pool of 'non-persons' considered disposable without any compunction within contemporary society, such that no moral injustice is seen to have been committed.[12]

Neither does it seem that we, as created beings, shall ever through our own 'bottom-up' capacities, achieve an Edenic paradise as exemplified in that 'Old Creation' narrative. That air of restlessness in today's cultural ambience is exemplified through moral dilemmas, for instance, in the seemingly unresolvable issues concerning abortion and the starkness of pro-life and pro-choice stances adopted by those on either side of the debate. Few people, it would seem, are aware that 200,000 abortions are performed annually in the UK, while across western Europe, UK, and North America, the annual total is 2 million.[13]

And so, to end-of-life matters. The aim here is to ensure that a proper moral status and dignity is extended to all those unable to mount their own defence including end-of-life frailty. Currently, within the UK, there is much clamouring for achieving an easeful end for those approaching death, in line with what already happens in certain jurisdictions in Europe and the USA. If euthanasia were to be similarly legalised, would there be a similar, dramatic rise in the deaths of the infirm or incapable, as is now evident from the statistical data for abortion practice? What kind of reasons would be offered in support of legalised killing? And how would attitudes change within the law, but over time?

Such issues I discuss in depth in comparison with UK abortion practice and how, during the 40-odd years of its operation, the Act's original stipulations have been bent with time. In Chapter 10 (Assisted Death: Legal Issues), I illustrate how those changes affecting the current use of the UK Abortion Act could be used as a source of wisdom in formulating legal approaches to assisted suicide and assisted death. At least, in my view, this would overcome some of the weaknesses in the proposal by Lord Falconer now being debated in the British House of Lords.[14] My proposals are given in extended fashion in this chapter.

In Chapter 11 (Assisted Death: Theological proposals), on the other hand, I come specifically to a theologically derived approach to assisted dying and death. Despite widespread 'debate', the factions are divided, rather like abortion, into pro-life and pro-death supporters. Doubtless all Western societies will ultimately come to have legalised assisted dying. Those provisions already enacted have not been influenced nor modulated to any great extent by profound, Christian-based thinking or argument. That worries me.

It follows whether our behaviour to those in end-of-life care is morally acceptable, and based on reasonable perceptions of our respect for, and towards, others. This seems necessary as many Christian people will be caught up in having to make these decisions and in feeling guilty having done so, particularly regarding the death of a loved one. In this chapter, I offer a theological basis for coming to terms with the terrible agonising guilt likely to accrue on many of these occasions. In doing that, I have dismissed the usual epithets such as 'made-in-God's-image', 'sanctity of the flesh', 'God gives and God takes', or that 'life is a gift'.

I base my argument on the life of Jesus, his compassion towards those in some kind of need, his having been made to wait while under arrest, and upon the cross his final sacrifice for the world. Those three phases can be aligned with the suffering, waiting, and ultimate sacrifice of bodies simply no longer capable of delivering, or physically sustaining, the kind of life their possessors would wish for. At the same time they could also be envisioned, analogously with Jesus, as states endowed with holiness, and thus sanctified. That being partly 'divinised' through a baptismal rebirth from above (Chapter 6: Theological Anthropology) and hence partially 'en-hypostatised' within the Godhead, provide the necessary background to a theology of sacrificial dying, when the body (as flesh) has no persisting value and could then, (in those exceptional circumstances), be done away with.

In summarising the thoughts, ideas and trends occupied through Chapters 7 -11, the issue comes to rest on current notions of what a person is. I favour, as most sensible people would, 'who-ness', which encapsulates each specific individual as the *one who is*, as enshrined through baptismal and eucharistic congress with the Godhead and which posits within that same Godhead, the foundations of human Dignity – for 'all sorts and conditions of men' (as the Book of Common Prayer puts it). Only in coming face-to-face with the Godhead will our true selves be revealed. This approach overrides the common, superficial application of 'what-ness', the latter failing in its attempt to capture the specific, unrepeatable features of each human being as unique, as seen through the eyes of the Godhead. What-ness is a useage which, because of its dependence solely on mere descriptive this-or-that physical and social attributes, or to material possessions, is unable to frame particular subjects in their unsubstitutable identity. It also overcomes those current trends, which exclude all the helpless at the extremes of life from the so-called 'moral community' of the supposedly fit and intellectually mobile people in society.

This brief, introductory survey should be sufficient to arm the reader with the scope of the material included in this book. It covers a wide-ranging overview of what it is to be human. Our evolutionary history is impressive, our acquisition of language perhaps being the key factor which permits abstract thought leading to our superior intelligence, technological advances, and the resulting cultural breadth of achievement. In this secular age, that account is sufficient for many people to live their lives on earth. In my view such a perspective is faulty, so that an additional theological approach gives substance to the inescapable thrust for 'transcendental' escape from mere creatureliness and its contingent horrors, and the distant bleak horizon encompassing death. In addition, the reason I think a theological approach is necessary is to counteract the extensive curriculum of disability as well as the evils, intrinsic and extrinsic, which can wreck lives, families and livelihoods. In our current world, we do not have to look far in order to garner the evidence. I am not convinced that secularist philosophies can totally accommodate such evil, which we see around us in the world at large.

It is the incarnational work of Christ which helps to give true meaning to those events, but which cannot always eradicate them from the contingent nature of earthly existence. However, through that redemptive sacrifice, these horrors acquire a new significance, since it demonstrates that the Godhead is not powerless to bring restitution, albeit in the world to come. That assessment is not mere pie-in-the-sky, as the relevant chapter on disability, for example, clearly demonstrates.

I do not wish to write a manifesto of despair or gloom, since there is much in life which, through its myriad achievements and accomplishments, can lift us to new heights. But we must remember, nevertheless, that for countless humans, life is a hard, unremitting grind, full of inequalities so often robbing them of the prospect of a life well-lived, imbued with love, enjoyed, and out of which some constructive usefulness seemingly emerges. That is why I am earnestly convinced that there must some kind of reparation for those who never made it on earth, such that there must be a world to come founded on a theistic base of love and ethical restoration.

But the 'world-to-come' could never be carbon-based, forged out of the cosmic cauldron of star-bursts as we understand it currently, and necessarily subject to the entropic influence of the Second Law of Thermodynamics. Furthermore, if we were to continue to exercise 'free-will' in the afterlife, this would inevitably incur the occasion of sin, unless we are reconstructed as mere divinely-activated puppets.[15] And arising out of the carbon-based pastoral of Isaiah (11:6-9), we would surely need leucotomised lions disabled from devouring the lambs lying aside them: but then, they would no longer be recognisable lions. I know of no one who has the ability to conceive how a future New Creation might be envisaged – a vista freed from the inevitable, contaminating thought-bias of carbon-based existence. The best we have come up with, so far, are spindly Martians with potato-shaped heads sprouting green hair. At least they are able to claim some kind of distinctiveness – if not notoriety.

On the other hand, to be in a holy, sanctified realm we, as resurrected humans, could only exist within the Godhead in order to maintain purity, although nobody has ideas as to how that would, or could, be accomplished.[16] Therefore, we must remain ignorant. Nevertheless, it is that sacred aspect of a future life which gives dignity to all – including the smitten, and which thus affords them reparation against mindless destruction upon earth, either through geophysical catastrophe, pandemic infectious disease, or man-made evils including war, abortion, infanticide, reckless euthanasia or other personally-directed assaults, verbal or physical.

There is much to learn about *being human*, as I recapitulate in Chapter 12 (On Being Human) and I hope this books articulates those universal concerns which define life, its contingent but frequently forgotten and overwhelming vicissitudes, and how it should be valued and lived, particularly with a far greater respect for the other person than is, at present, evident. There is, I believe, hope: yet the question remains how that hope can, and should be, configured – and realised in 'the end times' – that is, eschatologically.

* * *

1. Tallis R, 2011. *Aping Mankind: Neuromania, Darwinitis and the Misrepresentation of Humanity*. Durham: Acumen.
2. Scruton R, 2011. Neurononsense And The Soul. In: van Huyssteen J & Weibe EP (eds), *In Search Of Self*. Grand Rapids: Eerdmanns 338-356.
3. Marsh MN, 2010. *Out-of-Body & Near-Death Experiences: Brain-State Phenomena or Glimpses of Immortality?* Oxford: Oxford University Press 237.
4. Zizioulas J, 1975. Human Capacity and Incapacity. *Scottish Journal of Theology* 28: 401-447.
5. Zizioulas J, 2006. On Being Other. In: *Communion And Otherness: Further Studies In Personhood And The Church*, (ed. McPartlan P). London: Clark 37.
6. Gordijn B, 1999. The Troublesome Concept of the Person. *Theoretical Medicine & Bioethics* 20: 347-359.
7. Ford N, 1988. *When did I begin?* Cambridge: Cambridge University Press.

8. Warnock M, 1984. *Report of the Committee of Enquiry into Human Fertilisation and Embryology*. London: Her Majesty's Stationary Office, 1984.

9. Barnes C, 2008. Generating Change: Disability, Culture and Art. *Behinderung und Dritte Welt* 19: 4-13.

10. Keate G, 2015. One in three children has suffered violent attack: 385,000 assaults go unreported as crime becomes 'normal'. London: *The Times* newspaper, January 1st 1-2.

11. Editorial, 2015. The Other Pandemic. London: *The Times* newspaper, January 7th 26.

12. Smith Wesley J, 2000. *Culture of Death: The Assault on Medical Ethics in America*. New York: Encounter Books.

13. London: Office for National Statistics (ONS), 2010.

14. Marsh MN, 2015. 'Te Lucis Ante Terminum': A Perspective on Assisting Suicides. *Modern Believing* 56 (2), in press.

15. Berydaev D, 1936. *The Meaning of History*. London: Bles.

16. Fiddes P, 2000. *The Promised End: Eschatology In Theology And Literature*. Oxford: Blackwell Publishing.

Chapter 2

Physical (Evolutionary) Anthropology

The universe originated with what cosmologists call 'The Big Bang' approximately 13 billion years ago. Life, whether from unitary or multiple sources, arose on this planet about 400 million years ago. Approximately 5-6 million years ago, a split from our 'last common ancestor' gave rise to two subsequent lineages, the chimpanzees and another from which hominins (or 'pre-humans') would ultimately emerge. The origins of 'anatomically modern humans' from these pre-human hominins began around two million years ago, at which time the primitive chimpanzees also underwent another split to produce the additional miniature, or Bonobo, chimpanzee. Some ideas about our common heritage and cultural development derive from the finding of ancient skeletal remains and artefacts in various parts of Africa as well as in Eastern Asia and across Europe.

We may all be familiar vicariously with some of the names of these ancient predecessors of modern humans, including *Ardipithecus ramidus* – a female who lived over four million years ago, followed by several Autralopithecine/Homo species (*habilis, ergaster, erectus*) leading to *Homo heidelbergensis* which probably evolved within the last one million years. There followed *Homo neanderthalensis* (originating c.300,000 million years ago) and the primitive *Homo sapiens* (c.200,000 million years ago). Some skeletal finds have been dubbed with popular names, such as 'Lucy', 'Peking Man', and so on. For some considerable time, *Homo neanderthalensis* overlapped with other Homo species, but eventually died out around 30,000 years ago for reasons that are not entirely certain. DNA analysis reveals the presence of Neanderthal gene sequences within the modern human genome, indicative of cross-breeding and therefore presenting a completely new approach to our understanding of our ancestors' cultural behaviour.

As is clear from the literature, the way in which modern humans developed is by no means certain, since there are scanty examples from each presumed epoch, and fewer as we recede further back into history. Many hypotheses about origins derive from a few bone fragments, since complete skeletal remains are extremely rare. The recent discovery of the remains of a cranial face from a cave in Spain demonstrates very modern features, yet because there are no other skeletal remnants available for comparison, it is difficult to classify this individual, although its characteristics suggest an evolutionary place, perhaps, just before *Homo heidelbergensis*.

Therefore, we do not have a precise linear pathway through which our origins can be mapped: nevertheless, we should envision our evolutionary past as more reminiscent of a branched bush, rather than a straight linear tree trunk. In other words, we still have to determine who our parents, brothers and sisters are and hence, our direct grandparents and great-grandparents, but distinct from our aunties, uncles and cousins within the wider hominin 'table of affinities'. It now seems evident that several evolutionary hominin species (among the Australopithecines) overlapped and hence were walking the planet at the same time.[1] Importantly, they have all been extinguished within this very short evolutionary time period, save for *Homo sapiens* to which we belong.

Advances in laboratory techniques are allowing DNA samples to be retrieved from much older specimens. This permits cross-identification of species, and provides information on the evolutionary and mutational changes now detectable in protein structure (so-called 'molecular chronology'), thereby offering additional, ancillary approaches in determining the true patterns of evolutionary development. Ancient geological strata in Africa and Eurasia are rich in ancient fossils, and continuing discoveries are altering current perceptions of the pathways through which modern humankind is derived. What is certain, today, is that the view prevalent about 30 years ago – that this new (human-like) species began jumping out of trees on two legs (c.4 million years ago), walked into East Africa, developed into pre-humans (c.2 million years ago) and then migrated into Eurasia (c.1 million years ago) – is now defunct. The picture is far more complicated. Nevertheless, bipedalism was the earliest alteration which led to our appearance on this planet, even though walking in the beginning resembled a bandy-legged infant with inwardly pointing feet.

An example of the new finds that can bring seismic changes to our understanding of human evolutionary anthropology has come from the Malapa Caves, South Africa, 2008. Geographically these rock formations are dated to two million years ago, although the extensive skeletal remains were actually found in a pool or lake now submerged some 50 meters below land surface. How these early species got there is uncertain, and there were no associated animal bones: conversely the skeletal remains luckily were not disturbed by predatory activity. Two of several remains have been described, a boy about 12 years of age (MH1), and a young female.[2]

Primitive features included small stature and a small brain, the cranial vault estimated to be rather apelike with a volume of about four hundred and twenty cubic centimetres. Yet more 'advanced' characteristics included smaller teeth, the shape of the dental cusps, non-protuberant cheek bones, a prominent nose, long legs, and a pelvis whose contours matched those of *Homo erectus* (the latter also thought to have arisen two million years ago – the point at which ancient and modern humans begin to diverge). Despite those promising features, some experts have disagreed with the findings, arguing that there has to be a spectrum of changes associated with growth, age, and gender, so that the differences interpreted as 'more advanced' could,

23

possibly, merely be related to the developmental age of the skeleton examined.

The difficulties in assigning new finds to either existing, or to an entirely new species, and hence of having great relevance to epochal phases in human evolution, are clearly exemplified by these dilemmas. Another example of the complexities involved is seen in the tabulated comparisons of the Malapa skeletons with nine other specified archaic or pre-human types, with reference to the cranial vault (11 features analysed); facial skeleton (39); teeth (13) and 62 other structural data points. Yet this is not all.

A scan performed to very high resolution on the skull of MH1 while still containing all the internal time-accumulated debris (The European Synchroton Radiation Facility, Grenoble), revealed an external brain surface which shows 'later' humanoid features including the convolutions around the expanded frontal lobes, together with another prominent bulge (inferior frontal gyrus) related, in modern humans, to social activity and language. Indeed, given the age of these remains, it is quite possible that some semblance of speech had developed at this evolutionary time (see Chapter 6 concerning language acquisition).

These developments (and there is more to come from the huge deposit of bones in these caves) illustrate the complexities of the task as well as the excitement when the remains of almost complete, new skeletons are found. The data suggest again, not a direct linear descent, but a progressive, piecemeal or mosaic accretion of small changes, so that evolution seems to progress by an amalgam of very small advances exhibited by each of the extant fossils now categorised.

But that other side of the coin has recently been exemplified with a more refined examination of the remains of so-called 'Hobbit' man – a dwarf-like individual discovered in the Liang Bua Cave in Flores, Indonesia.[3] A modern evaluation, despite previous opposition, revealed that this person appears to have been suffering from Down's syndrome, and the shape of the skull and other bones is consistent with data deriving from contemporary people with this condition. That is, it does not represent another special branch of the hominin family, but seemingly a variant of Australo-melanesian *Homo sapiens* model which existed about seventy five to one hundred and twenty thousand years ago.[5]

Despite all that, it is remarkable how many species came into existence – and disappeared – during this relatively short evolutionary period of 5-6 million years.

Considering Early 'Human' Society

These introductory thoughts not only indicate the enormous complexities pertinent to clarifying our descent, but point to the required level of technical information necessary for understanding the process of inter-species anatomical comparisons between bones, skulls and teeth. However, for our purposes, we need not travel along those pathways. Yet there are additional approaches made by scientists, which have broadened the contours of our perceptions as to how these ancient peoples lived – and survived. Several approaches have been explored.

One intriguing avenue explores the ratios of stable isotopes of carbon (^{13}C and ^{12}C) in ancient bones and teeth. These give a fair estimate of the types of diets that were eaten at different times throughout evolutionary history, thereby giving important clues about the habitats of the different species. The C_4 pathway (favouring ^{12}C over ^{13}C content) characterises water-containing plants and grasses, compared with the C_3 pathway (favouring largely ^{13}C content) more typical of shrubs and trees. Those species doing better had a 65/35 mix C_4/C_3, whereas earlier species and those that succumbed tended to adhere to the harder C_3 diet. Nevertheless, it is surprising the extent to which earlier and more arboreal-inclined species showed a C_4/C_3 dietary bias. But it certainly seems true that increasing mobilisation allowed certain species to diversify and avoid abrupt changes in food availability that were due to shorter term climate changes.[4,5] That preferential change in diet not only reflects a bias towards C_4 plants, but possible access to some meat from animals ingesting a similar C_4-based diet, the ingestion of termites, use of root vegetables and tubers, or the consumption of brain and bone-marrow. This is borne up from evidence deriving from about this time, for crude stone implement usage on animal bones (for example, *Homo habilis*: approximately from 3-2 milllion years ago). The C_3 diet would have been rather hard, even gritty, and consistent with the degree of scoring identified by microscopy on the teeth of these individuals. The evidence also suggests greater mobility occasioning access towards a more open, pastoral and water-side environment, rather than one entirely arboreal.

Another fascinating approach has been geophysical, with the sampling of lake beds and analysis of the long, excavated cylindrical cores (thirty-feet in length) representing ~250,000 years accretion. These deposits are not subject to the disruptions influencing the layering of terrestrial landscapes, and thus give a better picture of climatic conditions over long periods of time, and which are added factors that would have shaped human evolutionary history. Every 1.5 inches of these sampled cores represents ~1,000 years. The cores reveal alternating dark and light patches, relevant to wet and dry seasons, which regularly alternate over a cycle of 23,000 years, being directly influenced by the earth's wobble. Therefore it is clear from that information alone that ancient man would have had to dodge the wet and dry bits in order to forage effectively. This is quite different from the earlier view that man arose to inhabit and walk through a continuously dry, warm savannah-type landscape.

But these alternating wet/dry seasons caused by planetary wobble had superimposed upon them longer-term seasons that were either warm, or interspersed by extended, frozen, ice-bound epochs. It seems that such a change occurred around 2.9 million years ago with the extinction of *Australopithecus afarensis* ('Lucy') and the appearance of two other species, one of which is known as *Homo habilis* because of the crude stone tools located near the skeletal remains. This species lasted for nearly 900,000 years, and without much apparent change in cultural features throughout this prolonged period. A further perturbation occurred around 1.9-1.6 million years ago. *Homo habilis* became extinct and there then emerged two more human-looking types: *Homo ergaster* and *Homo erectus*. The newer species were taller, had longer legs, and an increased brain size of around 750cc. Their tools reflected this change, being far better crafted, and bearing blades very carefully forged on both sides. The increased brain size demanded a greater calorie intake, a difficulty probably overcome by group foraging over greater distances. Yet there is little evidence that spears and accurate throwing were employed at this time, and there is scant evidence for widespread use of fire for cooking. According to Dunbar, laughter would have provided an important social cement contributing to a collective sense of well-being, as a behaviour which overtook individual grooming.

Such underlying climatic changes, it is now thought, were responsible for these remarkably rapid, evolutionary develop-ments. In other words, with dramatic climate changes, favoured foods become scarce. As a result, a species must either be able to adapt and brave the elements, or perish and become extinct. Environmental forces, manifested as abrupt changes in climatic conditions, favour those genes that initiate useful characteristics – such as increased and effective mobility, and a larger brain which facilitates greater awareness in coping with the environment. It should be noted that evolutionary history has seen five massive extinctions, all followed by bursts of new activ-ities and the appearance of completely new species or markedly better-adapted successors to those who existed previously.[6,7]

Considering Intermediate 'Human' Society

Despite their newly acquired physical properties, *Homo ergaster/erectus* lasted during the interval 1.8 – 0.5 million years ago. In turn, they were replaced by a more human-like species, *Homo heidelbergensis* (from 600,000 years ago) and then *Homo neanderthalensis*, from about 300,000 years ago. These types colonised Africa and Europe and reached Siberia (the so-called Denisovans). They had enlarged brains exceeding 1.200cc. They were of stocky build, and had prominent furrowed brows and flat, terraced skulls. They showed signs of environmental stress, since the skeletal remains reveal marked arthritis, skulls with penetrating injuries, and stress lines in their dentition, even among children, indicative of multiple infectious episodes.

Nevertheless, from this era we have the most cogent evidence for the use and control of fire. Although a contentious issue, cooking makes available by at least 50%, a more nutritious diet from heated food, especially meat.[8] They had spears, tipped with shaped stone arrow heads. From the bones associated with these skeletons, it follows that they hunted a variety of medium-sized herbivorous animals, and probably used the remains for other uses within their habitats. Studies of the carbon and nitrogen isotopes in their bone collagen also corroborate their meat-eating habit.[9]

The larger brains of these later hominins suggests that they were reasonably intelligent, although the shape of their skulls is different from that of modern humans. Their frontal lobes were not well developed, but there was a protrusion ('bunning') of the posterior areas, suggesting an enlarged visual cortex. Indeed, there does seem to be a correlation between an enhanced cortical visual system, and the possession of large eye sockets, both features having a direct correlation with smaller latitudes. This is because the light quality would have been somewhat poorer and rather more restricted in northern European territories as opposed to Mediterranean climates. That is because the days would in general have been shorter for longer periods, especially during the winter, late autumn and early spring, and because of the inclination of the earth to the sun, the light had to take its longer course through the atmosphere.

One problem which has intrigued anthropologists for decades is why all the Neanderthals, after inhabiting Afro-Eurasia for over one-third of a millennium, suddenly vanished. There have been many theories attempting explanation. The most likely reason is that they were over-run and out-smarted for food, fuel and shelter by the rapid increase in what are termed anatomically modern humans (AMH) even though both populations co-existed for 5,000 years. Detailed study of the remains of densely inhabited sites across south-western France across three epochs (35,000-40,000; 40,000-44,000; and 44,000-55,000 years ago), the earlier two periods largely occupied by Neanderthals, provided these data.[10]

The conclusions derived from the total number of sites occupied, their intensity of usage and occupation, and the extent of the geographic terrain occupied showed substantially more activity during the latter period, which involved AMH only. Based on calculations, it was concluded that AMH ultimately outnumbered the Neanderthals by a ratio of 10:1, thus displacing them to the extremes of these regions, an evacuation which gradually saw them become extinct. The ultimate factor helping their demise was the arrival of a rapid climatic event which saw the temperature drop by 10°C within a decade, around 40,000 years ago, with resulting icebergs in the Mediterranean Sea. The last Neanderthals occupied its northern littoral in caves distributed along the southern coast of Gibraltar and Spain.

Considering Late Human Society

Anatomically modern humans (AMH) arose around 200,000 years ago in Africa, replacing all trace of archaic hominids during the next 100,000 years, other than Neanderthals and Denisovans who still inhabited much of Eurasia. A major 'out-of-Africa' migration of early *Homo sapiens* took place around 70,000 years ago, with penetrations into the East and Australia by 40,000 years ago, by which time, as we saw above, the Neanderthals had become extinct. By 20,000 years ago, *Homo sapiens* had reached the American continent – and by foot.

We now have to consider what makes a community, and how the organisation of later human society differs from that of earlier hominid species. One very useful approach to this subject has been pioneered by Professor Robin Dunbar in Oxford, with other colleagues.[11] They had realised that working merely with bones-and-stones would not provide any further understandings about communities and their origins. With that came the additional realisation that it was essential to incorporate the work of other specialities in order to enhance this aim, from psychology, neuro-physiology to computer graphics and so on, thus giving rise to the intriguing concept of the 'social brain'. While it was known that the human brain had trebled in size through about seven million years of evolution, what additional insights could be drawn from neurological understandings of brain function?

First, cognitive abilities are important, but that predominance should not overshadow a more global view sensitive to other cultural achievement, as these authors rightly point out. Above, it was noted that the increased size of the Neanderthal brain was largely accounted for by its expansion of the posteriorly-located visual cortex, probably related to the need for improved vision in poorer (northerly) light conditions. Modern humans, on the other hand, evolved brains characterised by enormous frontal bossing due to the anatomical development of the frontal lobes and of the functional capabilities represented by the pre-frontal cortical areas.

Yet with pre-Heidelbergian species, we have to be a little more imaginative and realise that even a primitive stone tool simply does not appear from nowhere. It is necessary for the idea of using a roughly splintered stone for cutting flesh, bones, or fibrous roots to enter someone's head, and for the concept of its use to be realised and understood before the reality can take place. Following that, the further idea of shaping a stone on both sides developed, to obtain a sharper surface for improved cutting efficiency. Then somebody thought of attaching a very sharp flint-head to a long branch for use as a weapon to kill beasts. And moreover, these technologies were then copied, passed on to, and adopted by others thus becoming generally accessible to the whole community. Although chimpanzees use stones to break nuts, such developmental elaboration and 'transmissibility of culture' has never been observed with those animals: that is the difference. From all this, we can discern the existence of a theory of mind exceeding two orders of thought, or more, thus exceeding the capabilities of apes and monkeys. In other words, 'I know, that you know, that I...'.

What becomes clear is that the Neolithic revolution cannot be used as a starting point for understanding the trajectory of human evolution. It is necessary to return to earlier ages and assess competences, as illustrated above. This is not easy, since a significant change in brain size is by no means always, or even necessarily, correlated with either a noticeable alteration in cultural achievement or direction. In passing, the dilemma of increasing brain size brings obstetrical problems, in the pay-off between walking efficiency, with its necessity of a tilted pelvis and long legs, as balanced against the maximum neural development that would guarantee post-partum foetal survival. These considerations critically bring out the social needs for midwifery and how that was developed in relation to obstructed labour, and the high risk of both intra-uterine and maternal death rates through evolutionary time. While the record seems not to offer up fragments which could illuminate those pressing outcomes, the need (and even its occurrence) must not be forgotten.

Second, there is the important role of emotion, beyond mere cognitive events and, in my view, one of the most impressive outcomes of the social brain theory. It is this viewpoint which so incisively puts early human communities into context, as a

31

means of understanding their ethos and philosophies. We are all aware of the angry outbursts that are seen in troupes of chimpanzees. Co-operative behaviour has been observed among these animals, but when personal needs are threatened, then those personal demands are protected and fought over. Is it not possible that small groups of early Australopithecines might also have behaved in similar ways? Following the Heidelbergs, when brain size had significantly increased, it is possible that frontal lobe connectivity between the 'top-down' controlling pre-frontal cortex and the limbic system might then have started to have a modulating effect on that type of aggressive behaviour, although not eradicating it completely. Indeed, we see such sporadic uncontrolled bursts of temper today, but behaviour, in general, is a little more restrained. And like us, they may have entertained forms of punishment and strict criteria of fairness.

Third, there is another significant factor that plays a role here, and that is fire. Good archaeological evidence indicates that fire had become widely used and controlled by 400,000 years ago. This was a great advance in the cooking of raw meat and hard vegetables, rendering them far more digestible and opening up greater calorific value by up to 50-fold. But fire has an important social effect, as we know from our personal experiences today: there is nothing so dreamy and consoling as the idea of eating and then sitting round the camp-fire. A fire also prolongs the daytime period. In a highly significant recent paper, conversations were recorded from !Kung bushmen in Africa.[12] While daytime fireside talk was fairly trivial, almost similar to the banality of Western-style coffee mornings, the tenor of the talk through the evening was completely different. At those times, community spirit was at the fore, with recollections of past-times, deceased tribal forebears, and thoughts encompassing aspirations towards the future, and depictions of the 'big picture'. We should be in no doubt about the role fire would have played in inter-personal bonding, and in forging community togetherness.

Here the social community would have comprised of around 150 individuals, as is commonly the case throughout society today (mediaeval villages; a military 'company'; church congregations; Christmas card lists, etc). Additional bonding behaviour would have included music, laughter, dancing and rhythm. Rhythm, (including synchronised music), plays an important role in the efficient binding of various kinds of activity, including the flocking of birds and shoal dynamics among fish, in addition to other mutually-interacting complex systems.[13]

In more intimate settings, the number of close relationships seems to be set at 3-7, as would obtain within family settings. How did families, as we know them today, arise? This is linked to questions regarding the origins of monogamy and the estimated epoch, during evolution, when such lifelong relationships began to take shape: it might reasonably be surmised that the increased size in the hominin brain would be a major factor, and there is a close correlation between brain size and group community size. The evidence reveals that early human-like species were not much different from apes or monkeys. In all those species, which are polygamous, the size of males greatly exceeds females, brain size is small, and females live a harem-type existence. This is also confirmed by a small second digit (index finger, relative the fourth (or ring) finger), or low 2D:4D ratio. A similarly low ratio is demonstrable throughout all early humanoid species, and only equalises in post-Heidelbergian individuals, where the size of males to females also approximates equality.

From that, we could most sensibly conclude that monogamy occurred relatively late in evolution, followed by bi-parental child rearing, with a consequent reduction in infanticide by other males wishing to mate with that particular female. Monogamy is, however, a socially exclusive – as well as divisive – departure, and considerably alters the way we now live, both as pairs, and as members of the local community. Human weaning is comparatively rapid whereas child growth is extremely protracted, and employs considerable inputs as a trade-off for continued brain growth and pre-adult social learning.[14] Another curious feature unique to humans is the extended longevity of menopausal women, who contribute to the rearing and feeding of their daughters' children, while enhancing the mother's availability for further fertilisation.[15]

The concept of the social brain embraces most definitely the cultural achievements noticeable in the settings of ancient mankind. One further aspect of early man's behaviour, seen from the context of this book, is burial practice. The earliest site found so far (in Israel), is 130 million years old, suggesting a widespread practice. Moreover, burial is less strongly related to the control of fire than to bead technology. These burials clearly indicate considerable thought about those who had died, where they might now be, and what their futures might have entailed.

Finally, another endearing feature is that the materials employed in bead making came from sources some distance away from the community's location, raising the notion of time and energy deployed for their gathering. These ornaments, including shell, ivory, amber, bone, jet and other plant stuffs, have implications for a sense of self and the projection of that self into the community, of intimacy, self-awareness, and a focus on the body of the wearer and concepts of beauty. They could also point to dyadic relationships and the emotional cost, desires and loving feelings of giving and receiving, which such unions imply. Not only were these precious objects worn around the neck and wrist, they were additionally attached to clothing as featured adornments – as the tomb depictions suggest.

Nor should we forget the most impressive cave drawings which at Chauvet appear to be at least 30,000 - 35,000 years old – compared with the cave engraving from Gibraltar made by Neanderthals and dated to ~39,000 years ago – are almost certainly contrived by more modern individuals.[16] The depiction of power generated by herds of rushing beasts is stunning. Modern morphometric analyses reveal that the animals displayed were not dragged into caves and drawn, but remembered from what had been closely observed in the wild, so that cognitively-perceived details, such as the powerful muscular fore-quarters of these animals, and the thrust of their hind legs, were exquisitely conceptualised and accentuated in the drawings rendered.[17,18]

This has been a brief foray into the complex world of evolutionary thought. From that background, we can proceed to think about other contributory attributes of humankind – genes, conscious-awareness, and language. In passing, it should not be forgotten that the idea of a 'social brain' helps considerably in welding many findings, ideas, and concepts into a whole, which, although subject to future change as new finds are revealed, nevertheless provides a firm basis on which to evaluate all that is known to date about this intriguing story.

* * *

1. Dunbar R, 2014. *Human Evolution*. London: Pelican.
2. Berger L, de Ruiter D, Churchill S et al, 2010. *Australopicethus sediba*: a new species of Homo-like Australopith from South Africa. *Science* 328: 195-204.
3. Henneberg M, Eckhardt R, Chavanes S, Hssu K, 2014. Evolved developmental homeostasis disturbed in LB1 from Flores, Indonesia, denotes Down's Syndrome and not diagnostic traits of the invalid species Homo *floresiensis*. *Proceedings of the National Academy of Sciences [USA]* 111: 11967-11972.
4. Choi C, 2013. Early human diets. *Proceedings of the National Academy of Sciences [USA]* 110: 10466.
5. Cerling T, Manthi F, Mbua E, et al, 2013. Stable isotope-based diet reconstructions of Turkana Basin hominins. *PNAS*

[USA] 110: 10501-10506.

6. Anton S, Potts R, Aiello L, 2014. Evolution of early *Homo*: an integrated biological perspective. *Science* 344: 1236828-1236828-13.

7. de Menocal P, 2014. Where we came from. *Scientific American* 311(3): 33-37.

8. Wrangham R, 2007. The Cooking Enigma. In: Pasternak C (ed), *What Makes Us Human?* Oxford: OneWorld 182.

9. Richards M, Trinkaus E, 2009. Isotopic evidence for the diets of European Neanderthals and early modern humans. *PNAS [USA]* 106: 16034-16039.

10. Mellars P, French J, 2011. Tenfold Population Increase in Western Europe at the Neanderthal-to-Modern Human Transition. *Science* 333: 623-627.

11. Gowlett J, Gamble C, Dunbar R, 2014. Human evolution and the archaeology of the social brain. *Current Anthropology* 53: 693-722.

12. Wiessner P, 2014. Embers of society: firelight talk among the Ju/'hoansi Bushmen. www.pnas.org/cgi/doi/10.1073/pnas.1404212111.

13. Hennig, H, 2014. Synchronization in human musical rhythms and mutually interacting complex systems. *PNAS [USA]* 111: 12974-12979.

14. Kuzawa C, Chugani H, Grossman L, et al, 2014. Metabolic costs and evolutionary implications of human brain development. *PNAS [USA]* 111: 13010-13015.

15. Hawkes K, O'Connell J, Jones N et al, 1998. Grandmothering, menopause, and the evolution of human life histories. *PNAS [USA]* 95: 1336-1339.

16. Rodriguz-Vidal J, d'Errico F, pacheo G, et al, 2014. A rock engraving made by Neanderthals in Gibraltar. *PNAS [USA]* 111: 13301-13306.

17. Biederman I, Kim J, 2008. 17,000 years of depicting the junction of two smooth shapes. *Perception* 37: 161-164.

18. Cheyne J, Meschino L, Smilek D, 2009. Caricature and contrast in the Upper Palaeolithic: morphometric evidence from cave art. *Perception* 38: 100-108.

Chapter 3

Genes, Genetics & Epigenetics

It has not escaped our notice that the [base] *pairing we postulated immediately suggests a possible copying mechanism for the genetic material.*

So ends, with (part) characteristic British understatement, one of the most famous papers, perhaps, in biological science. It was written by Francis Crick and his junior colleague, James Watson (an American citizen).[1] They proposed, in the journal *Nature*, a basis for the double helical structure of DNA. The success of their model lay in incorporating the bases inside the paired helices, the latter comprising a molecular backbone of ribose sugars and phosphates. The bases guanine (G), cytosine (C), adenine (A) and thymine (T) are always specifically paired: G-C; A-T. In nucleic acids, ribose sugars are pentoses, linked in 5-3 fashion. That means that at the top of the helix, an unattached 5 position is free, and similarly at the other end, a 3 position.[2]

The helical structure of DNA underpins the two essentials needed for life (a) replication and (b) the configuring of the proteins which contribute to bodily structure and function. The human genome contains approximately 3,000,000,000 base pairs broken up into 24 linear stretches of variable length, comprising each of the 48 chromosomes. Of these, 46 are termed autosomes, the remainder comprising the X and Y chromosomes. Every female has two X chromosomes (46XX), and males one X and one Y (46XY). As far as we know, there are ~20,000 coding genes, that is, they contribute to the formation of body components. The non-coding component (95% of the genome), has no specific function – as yet.

When a body cell divides (or undergoes 'mitosis'), the double DNA helix of each chromosome separates, as Watson and Crick postulated, each half being pulled apart by a flat web of contractile proteins (the so-called mitotic spindle). Once all the divided chromosomes are separated, the existing cell membrane inverts thereby producing two new identical daughter cells. However, the production of ova and sperm cells is different, and proceeds by a different mechanism to ensure that each germ cell contains only one half the total number of chromosomes. Thus all ova contain only 23 autosomes, plus an X. Sperm also contain 23 autosomes and either an X, or Y, sex chromosome. Thus, when an ovum is fertilised, the full complement of chromosomes is re-established. That, briefly, summarises how one aspect of replication works.

The manner in which the bases are aligned on each chromosome determines the structure of all the proteins necessary for body function. During the process, a small section of the double helix must be separated in order to expose the gene segment, which can then be copied, or 'transcribed' in technical language.

The 'genetic code' is dependent on the sequence of bases along each helix of DNA, a 'codon' consisting of three adjacent bases specifying one particular amino acid. The 'gene' sequence to be transcribed is termed an 'open reading frame' (ORF), and is read from left to right, from 5' to 3'.[2] There is a 'start' codon (eg. 5'- - AUG- - 3') which indicates where transcription should commence and 'stop' codons (eg. 5'- - UGA- - 3') to signify the length of intervening DNA to be read. If the initiation of transcription is faulty, for example, then the wrong sequence of codons will be transcribed, resulting in a non-functional protein and the wrecking of the metabolic process(es) in which that protein should be engaged. Conversely, if a mutation has occurred, then the changed base at that position (for example, a G for a T) will be incorrect, which may result in a faulty protein, thus disabling the relevant metabolic process.

The molecular processes involved in reading a gene and

transcribing it are extremely complicated. This is referred to as the Central Dogma meaning, generally, that DNA→RNA→ protein. In these events, RNA (ribonucleic acid) plays an important role, of which there are three major (classical) types: (a) RNA polymerase which actually reads the DNA and makes an RNA transcript or template which passes from the nucleus to the cytoplasm of the cell (b) ribosomes, huge molecular RNA-protein complexes, which then translate that RNA template ('messenger' or mRNA) into protein, by means of (c) transfer RNA (tRNA) which carries one amino acid, appropriate to the next (triple) codon on the mRNA template to its correct place into the enlarging peptide chain.

It is natural to think of DNA as a linear assembly of base-pairs (BP) in which the genes are somehow embedded. But the $3x10^9$ BP, if written out in 12pt Times New Roman, would extend over 5,000km (Montreal to London). The combined double helices are crammed into the nuclear space (~520μm^3), enclosed by other proteins termed histones, of which there are four types, each represented twice. The helix is wound around the histones, each complex representing a 'nucleosome'; the latter further packed into 'chromatin' fibres which are further condensed into chromosomes.

One might also think that all four bases were equally represented, yet special laboratory techniques indicate that approximately 60% of the genome comprises A+T rich regions. Furthermore, if the DNA is chemically broken up and fractionated, three fractions comprise G+C regions, of which one representing only 3% of the genome, contains >25% actual genes. In any differentiated cell type, most of the genome is silenced. Apart from genes required for structure, repair and energy, the remaining genes contribute to its continuing differentiated state, whether it is a muscle cell, pancreatic cell, or one or other type of neurone, and so on.

Thinking About Genomes

The genome describes the entire DNA material comprising any species of animal or plant. Although this is the 'book of life', as President Clinton was made to announce when the first completed human genome was sequenced, these complete genomes are, perhaps, not the last word, as might be first thought. In comparison with gorillas, chimpanzees, gibbons, there is not a lot of difference. In fact, our nearest evolutionary cousins, the chimpanzees, have around 95-98% genetic homology with humans (depending precisely on how those similarities or differences are expressed) raising the intriguing question as to why, phenotypically (meaning body shape, cognition, and culture), we are so apart. In addition, since the molecular configuration of each base is different, the three-dimensional structure of DNA is not wholly uniform, varying with each 'gene' segment: this aspect of DNA also conveys important information.[3]

Fundamentally, therefore, remains the question whether humans are unique and in what way, genetically, that uniqueness could have arisen. Of course, as many have pointed out, even if chimpanzees have >95% homology with humans, we are left with a <5% genomic difference. From that, it could be insisted that the difference could be manifested in several different genes, which could be specific – even unique – to humans alone.

If we think historically for a moment, that concept was in fact the cause of great fury during the meeting of the British Association in the Natural History Museum, Oxford, on June 30, 1860. Of course, the concept of heredity as the process of handing on body characteristics was understood, but not reckoned in terms of genomic sequences, genes, mutations, or polymorphisms (mutational differences) among each family of inherited genes. Indeed, it would be several decades hence before Gregor Mendel would call attention to the supposed 'agents' of transmissible heredity, through his monastic sweet pea experiments.

The issue at stake in Oxford on that fateful afternoon was Darwin's first edition of *The Descent of Man*, in which the inevitable conclusion had to be faced. Humans are not creatures designed and put on earth specifically by God, as the crowning pinnacle of all creation. No, far from it. We are descended from the apes, as Thomas Henry Huxley is so famously reported to have emphasised against Samuel Wilberforce, then Lord Bishop of Oxford.[4] Huxley continued that he would prefer to be descended from a miserable ape as [his] grandfather, rather than [from the forebears of] a man who would crushingly attempt to defeat a sound argument. After this riotous Victorian altercation, it is not entirely clear whether Lady Brewster actually did faint and neither, indeed, whether she needed reviving with smelling salts.

Nevertheless, it probably did result in some hurried revisions of pre-existing sermons extolling man being made in God's image, being in dominion over the animals, and so forth. But the major dogma – that the church's office is to continue dictating to scientists – was defeated. From now on, the idea of 'truth' would mean something different, and not based on superstitious idolatry, a view that still persists today. I am by no means certain that the current 'Science/Religion' debate has regained any of the ground lost, or even silenced the agnostics.

Of further interest in our considering the relationship between the genome and of being a human, was the later rise of 'eugenics' advocated strenuously by Francis Galton (1822-1911).[5] There was a widespread feeling both in England as well as the United States, that the nobler stock was being diluted out of existence by the lower classes and immigrants. Galton protested that any animal would want to be healthy rather than sick, vigorous than weak, and well-fitted than ill-fitted for life on earth. The Greek 'eu' meant good, having an endowed balance of all the right genetically-conditioned qualities, a balance influenced for humans by a good (that is, Eton-based) education. Reading between the lines, this meant coming from high quality breeding stock and passing on those highly principled attributes into future generations. It was that basis which was seen as necessary for raising and maintaining the domestic, social, and political tone – 'health, energy, ability, manliness, and courteous disposition' (sic) – and which society would be likely to refuse to recognise in criminals and others deemed undesirable. These principles, Galton stressed, should be inculcated into society, seen as a subject whose practical development needed social consideration, and ready to be introduced into the national consciousness.

This idea of dealing with the weakest members of society, despite opposition, caught on, and both the Rockefeller Foundation and Carnegie Institute funded activity in America, although it is wrong to conclude that this engendered Germany's espousing its Nazi programme of mass extermination and the creation of a pure, Aryan race.[6] Indeed there was much pre-war cooperation between Germany and the USA, and it may well be that the 1933 Law for the Prevention of Hereditarily Defective Offspring reflected the earlier legalisation of sterilisation in America. Indeed, the sterilisation of the feeble-minded continued in America until 1970. Of course, once the horrors of Auschwitz were revealed, and the Nuremberg trials had netted most of the major regime culprits, interest focussed on a family of nations initially spearheaded by the League of Nations. But in addition, the need for legalised human rights was seen to be necessary, as a counter to unbridled state interference in the personal lives of human subjects, whatever their status.

But to return: are humans unique by virtue of the creation of novel, evolved genes? The answer is that humans do have some differing, or altered, genes from other species, and present state-of-the-art techniques are continuing to reveal those differences. The differences are uncertain, since the necessary large-scale comparisons are not possible at present because there are so few complete genomes between our related animal cousins and humans to permit their precise identification. Furthermore, there is no direct relationship between a newly evolved, human-lineage specific (HLS) gene and even its phenotypic expression in any one individual. Positive changes can be identified, such as our brain growth trajectory, and the increased size of the brain – almost 50 per cent of HLS genes are concerned with brain growth and function. In addition, there is the loss of body hair and our increased density of sweat glands; the opposition of the thumb and facilitation of fine movements and dexterity; the descent of the larynx and development of muscles for articulation permitting the evolution of speech, and of language.

Of course, it might be insisted that those phenotypic changes are obvious. True, but they are now supported by knowledge of the genetic changes that underpin those progressive, yet momentous changes in body structure and function. The massive evolutionary development of the frontal lobes, and the widened controlling influences thereby exerted on cognitive performance and emotional responses in other parts of the brain, is likewise undergirded by many newly configured genes.[7] Given that there are only four bases, how could such variations be brought about?

It has been shown that numerous possibilities include chromosomal restructuring; single base pair changes; so-called insertions and deletions ('indels') of variable size; and the number of gene copies. An additional factor is the presence of 'transposons' and 'retro-transposons' throughout the genome – probably remnants of previously incorporated viruses from as far back as two million years ago. These are operative in the brain and may have contributed to its enormously complex structure that may have begun to originate from that era in evolutionary history. These replicating pieces of DNA/RNA are, in part, responsible for the fact that monozygous twins can never be 'identical' (despite the appeals by the media).

Another approach has been to incorporate presumptive new HLS genes into mice. From that model, it has been possible to infer the role of a specific gene which historically underwent accelerated evolution related to hand and foot development, and hence was crucial for stone implement usage, other finer tool-making activities, an improved gait with a forward-facing large toe, and bi-pedalism. Finally, in this review of novel genetic mechanisms, it seems that disease has also altered the human genome. Humans exhibit enhanced lymphocyte activity against pathogens, conferring on them the propensity for allergic phenomena and allied atopic conditions, and 'autoimmune' conditions in which bodily surveillance of what is 'foreign' turns on intrinsic structures causing them to be damaged and leading to varied forms of chronic ill health (pernicious anaemia, type 1 diabetes, gluten sensitivity) and even malignancies, such as chronic lymphocytic leukaemia.[8,9] All pandemics identified through our past have, in their wake, skewed the genome: leprosy, the Black Death, the massive influenza outbreak of 1918, AIDS – and now Ebola virus. Interestingly, in all these events, individuals have been found who were apparently 'resistant' to the disease – indicating that conventional Darwinian rules concerning hereditable principles are not entirely correct, and that additional epigenetic factors also modulate the genome rather than solely DNA mutational changes.

Epigenetics

Readers may already be feeling somewhat breathless after that tour of genes and genomes. Nevertheless, it has merely been a rather superficial review (bearing in mind those totally unacquainted with this enormous and rapidly growing area of bioscience), and a tour not given to any in-depth exploration of the molecular biological processes involved, nor to their fullest implications.

Hopefully, it will have been learned that genes are not hewn in stone: neither do they fulfil the pretentious epithet of being 'selfish'. Indeed, it will have been realised that genes can be altered by a variety of processes not necessarily conforming to strict Darwinian principles. That is, genes do not reign supreme, but are additionally influenced by intrinsic and external environmental factors. Those several influences on the genome allude to the growing science of epigenetics. Epigenetic outcomes may be transmitted in non-Darwinian fashion in the reproductive cells (sperm and ova), an aspect of their physiology which will be encountered with the later chapter dealing with the human embryo-foetus. It is important to be briefly acquainted with epigenetics, and the major influences now recognised within that field.

* * *

1. Watson J, Crick F, 1953. Molecular Structure of Nucleic Acids: A Structure for Deoxynucleic Acid. *Nature* 171: 737-738.
2. Conventionally, the helix is depicted horizontally as a linear structure, with the pentose free-5 position (designated 5') to the left hand side, and the unattached 3 position at the other end (3'), on the right:

    5' ......CCGATACGTTGCA .... 3'

    | | | | | | | | | | | | |

    3' ......GGCTATGCAACGT ....5'

    The complementary helix is not always shown, as its structure can immediately be inferred: mutational and other acquired changes, however, may need to be demonstrated more fully.

    Pentose sugars are 5-ringed molecules, each position, by convention, numbered.
3. Lanctot C, Cheutin T, Cremer M, Cavalli G, Cremer T, 2007. Dynamic genome architecture in the nuclear space: regulation of gene expression in three dimensions. *Nature Reviews Genetics* 8: 104-115.

4. Lucas JR, 1979. Wilberforce and Huxley: A Legendary Encounter. *The Historical Journal* 22: 313-330.
5. Galton F, 1904. Eugenics: Its Definition, Scope, and Aims. *American Journal of Sociology* 10: 1-25.
6. Allen GE, 2004. Was Nazi eugenics created in the US? *EMBO Reports* 5: 451-2.
7. Zhang Y, Lanback P, Vibranovski M, Long M, 2011. Accelerated recruitment of new brain development genes into the human genome. *PLOS Biology* e1001179.
8. O'Bleness M, Searles V, Varki A, Gagneux P, Sikela J, 2012. Evolution of genetic and genomic features unique to the human lineage. *Nature Reviews Genetics* 13: 853-866.
9. Chen S, Krinsky B, Long M, 2013. New genes as drivers of phenotypic evolution. *Nature Reviews Genetics* 14: 645-660.

Chapter 4

On Being Conscious

What comes into our minds when thinking about consciousness? Is it, to quote a lesser-known remark (made in completely different circumstances) by Sir Winston Churchill – 'a riddle wrapped in a mystery inside an enigma'?

In piercing the gloom, I prefer the phrase 'being conscious' – indicating that consciousness is not an additional appendage of higher living individuals. Likewise, John Zizioulas, the Orthodox theologian, reminds us that as humans, we don't just have life attached to our bodies: we *are* living beings.[1] In a similar vein, we are *conscious beings – being conscious*.

Being conscious, viewed as having full conscious-awareness of ourselves and the environment we daily inhabit is the essence of living – for humans and other animals, in contrast to being 'non-conscious', or 'unconscious' – for example, coming round from a faint, having nitrous oxide during childbirth, being 'knocked out' in the boxing ring, or recovering from a general anaesthetic. Viewed in that way, consciousness becomes a little more tractable and understandable. Neither should we forget our innate ability to render ourselves 'out of this world' in popular speech, every night as we fall asleep and awaken the next morning. That points to mechanisms within the brain capable of bringing about rapid alterations in levels of consciousness, or what neurophysiologists call 'state control'. Furthermore, notice how the switchover can be instantaneous – one moment we are 'out for the count', and the next, fully 'with it'.

In proceeding, my aim here is to explore some thoughts about the origins of conscious-awareness through evolution, and during foetal growth and development. Then I consider various subconscious states, in order to evince more meaning as to what being conscious is like, followed by the viewpoint that for most of our lives, we are able to function on 'automatic pilot' – we can do a lot of things without much thought or attention, but based on previous training, experience, memory. That raises further questions as to why we are conscious creatures, and thus why we need conscious-awareness.

The Evolutionary Origins of Consciousness

To begin: where does consciousness arise in the evolutionary scheme? Viruses and bacteria are presumed not to be conscious, yet who can doubt their continuing successful life-histories, in repeating the same old thing – and presumably in perpetuity – over billions of years in evolutionary time. They are still very much alive and evolving – yet 'they neither spin nor weave'. David Chalmers suggests that a thermostat might 'consciously' respond to changes in temperature.[2] But suppose, could a thermostat switch on at 32.5 rather than 30 degrees when having a bad 'hair' day and perhaps wanting to be rather naughty?

Are wasps conscious? Imagine how they endlessly and repetitively attempt to escape from a window – a behavioural pattern, I suggest, more exemplary of automata – neither understanding, nor even beginning to comprehend the presence and nature of glass, failing to learn that freedom requires an open window. Then there's the dance of the honeybee, revealing the approximate location and size of a newly discovered source of pollen. Is that a lesson in informational aerobics, or a pure example of inbuilt instinct?

Within the invertebrate/vertebrate borderland, it seems likely that octopi are conscious, are able to discriminate between varied environmental stimuli, to remember, and to learn from previous experiences.[3] Octopi have vertebrate-like eyes with retina and lens, but the presentation of pieces of meat with simultaneous electric shocks reveals their ability to 'memorise' the event and refrain from similar future indiscretions. This might, however, entail some reflex learning.

Moving forwards, I suspect that all vertebrates, to varied extents, are conscious beings.[4,5] Nevertheless, we must always be alert to instinctive traits and capacities, as exemplified by pet dogs, and not interpret their responses anthropomorphically rather than as automatic or reflex-determined behaviours. Fish demonstrate memory for new experiences learned, although they have few additional cognitive abilities.[6] Sheep remember human faces for two years, while ewes discriminate between the more attractive of two males.[7,8]

On the other hand, we recognise that apes and chimpanzees are conscious beings and capable of determined action and thought. Much accumulated evidence continues to underpin that assertion.[9] In tracing the evolution of the vertebrate brain, we can see its structural patterns already established in various pre-vertebrate creatures. Those inhabiting such a pre-vertebrate niche may have some degree of conscious-awareness, although these are contentious issues.[10,11,12,13] Clearly, as with babies, it is not always easy to infer that conscious understanding is present without some kind of separate output, either a motor response, or a verbal report.

Next, I want to split the idea of being conscious into two parts, the difference lying between *awakening*, in comparison with *attention towards*, or an *alert focus on*, specific environmental details. These ideas could relate to foetal brain development.

The Foetal Brain & Origins of Consciousness

When can we say that a foetus is a being with consciousness? There are few systematic studies, hence empirical data are scarce, and yet some careful thinking can be useful.

First, when doing my obstetrical training in Oxford, parents never seemed to fear that their offspring would not be conscious, but rather whether there was possibility of developmental abnormality. Again, parents might compare their offspring with other babies in terms of attractiveness, future potential, or intelligence – but they would never regard their baby's consciousness to be better, superior, or inferior to that of babies in adjacent cots.

Second, are babies conscious at birth? We all know that the new individual begins as a fertilised ovum, a single cell which obviously is not conscious (Chapter 7). But forty weeks later, that cell will have expanded into a new baby comprising around three billion cells constituted by ~250 differentiated cell types, and ready to enter the world.

But is a newborn baby a conscious being? If it were argued that a newborn baby is not a conscious being, we would logically be forced (a) to declare when, in its post-natal development, a baby does become a conscious being, and (b) to show how, and by whatever neurophysiological or psychological criteria, that change had now come about. If that were the case, it might be reasonable to assign to that intervening period some kind of appropriate descriptive terminology. That, I suggest, would not be a very promising line of enquiry: and it is a problem for those who might wish to adhere to such a position. It nevertheless raises the important question as to how we attribute conscious states to others, humans or animals, and by what criteria – especially (as we noted above) in absence of motor, verbal or written reports.

Third, if we assert that babies are conscious at birth, at what point does consciousness arise *in utero*? Furthermore, can we identify its so-called 'neurological correlates'? For example, we know that infants surgically delivered 'early' for clinical reasons at 36, or even earlier at 32 weeks, are conscious. But before 25 weeks, although survival is possible and the foetus is alive, is it also conscious?

The foetus has been shown to recognise its mother's voice and the characteristic prosody of her native language in comparison with strangers or foreign languages.[14,15,16] Furthermore the mother's voice is markedly amplified within the fluid environment of the womb. Thus, in immediately recognising mother's voice after delivery, the newborn is able to correlate her characteristic vocal characteristics with her facial appearance, which it does not see until after birth. This emphasises the important principle of continuity of conscious-being between foetus and newborn as it leaves the womb. Admittedly, this suggests that consciousness could arise as early as 20 weeks gestation. That is, sufficient neural structure becomes available to support consciousness, and by that I mean, *basic consciousness*. Before that time, there wasn't any apparent consciousness (at least as discernible with current technologies, or demonstrated by appropriate research), and now, at some point during foetal development, there is. Here, then, we can make distinctions between, firstly basic consciousness or wakefulness, and secondly arousal or alertness to internal body and external environmental signals.

The best analogy is of a TV set on 'stand-by': there is a current (or life) which illuminates the red pilot light – the wiring circuitry is 'awakened' (arousal). But there are no programmes imposed upon the set when on 'stand-by'. Analogically, there is little or no cognitive alertness, or focussed attention to specific external environmental signals.

That initial, basic 'wakefulness' or arousal needs to be compared with progressive foetal cognitive-awareness as birth approaches. The latter includes recognition of the maternal voice; responses to music played during the prenatal period; smells, touch, a sense of movement and of body-image; and the sensation of pain.[10] Here I think, additional detailed studies on the foetus and the immediately new born baby and beyond would have vital importance in determining (a) when basic consciousness arises for the first time, and (b) how the content of basal consciousness is expanded during the middle to the later stages of pregnancy and beyond, throughout the post-natal period. One important determinant is that of thalamo-cortical connections which conduct incoming sensory stimuli forwards to the cerebral cortex, and thence conscious-awareness. That connection occurs at around 25 weeks gestation.

The progressive assimilation of new cognitive and affective ways of behaving is analogous to downloading more programmes onto basic conscious arousal or wakefulness, a process through which a 'mind' is progressively acquired. It is through the downloading and tuning-in of more programmes (so to speak) through which the gradual and progressive refinements of the sensory inputs of vision, hearing, touch, taste, and olfaction are developed, all dependent on thalamo-cortical projections into conscious-awareness, and then correlated to motor activity by each person. Accompanying this development is the building up of effective memory systems: the acquisition of language, and the setting down of emotional or affective characteristics, and a dawning perception of relationship – at first, to the infant's mother, then to family, subsequently towards society and to the local environment – thence towards a fully-developed, mature adult mind. As these aspects of personhood continue to evolve through time, the growing individual develops the ability to think and adopts certain points of view. Through all these processes from intra-uterine life onwards, basic un-focussed, raw consciousness – or arousal – gives rise to the subsequent development of finely-honed systems of alertness – or of an attentional conscious-awareness which is progressively oriented to full intellectual development, and with it, the possibility of inter-personal relationships, the gradual perception of the transcendental sphere, and even a sense of the divine.

Further intriguing, if not paradoxical, aspects to that view of a developing mind arise: (a) from an evolutionary perspective, complete maturational growth of the brain requires a period of about 25 years, occupying at least one-third total life expectancy based on current-day statistics; (b) like all higher vertebrates, we are over-encephalized: relative to body size we are endowed with very large brains; while (c) a big brain puts demands on a mother who walks upright and whose pelvic outlet is limited. Thus, at birth, brain size is only ~25% that of adults, subsequent brain growth and the ensuing time required for brain maturation needing a considerable period of parental oversight in securing the long drawn-out acquisition of a trained, organised mind (Chapter 2).

These three aspects concerning brain size, its foetal development, and post-natal growth pose a costly biological adventure. How could basic consciousness ('awakening'), which is a non-physical 'mental' entity lacking a specific genetic basis, have been progressively selected for within the evolutionary process? Those thoughts are intriguing. As Raymond Tallis rather contentiously puts it – 'consciousness is an embarrassment for any truly, ardent Darwinian'.[17] And that, I think, is an extremely important observation. Consciousness (also regarded as conscious-awareness), viewed as awakening and subsequent alertness to the internal milieu and external environment, emerges in each individual once sufficient neurological kit comes into being, and it is underwritten by some very complex, integrative machinery: a brain. What kind of adventure was this – and why did it take place?

States of Not Being Conscious

While we temporarily set those questions aside, I now want to think briefly about certain 'unconscious' states. As this is an enormous field, I shall not deal in detail with every aspect.

However, in passing, we should briefly consider certain heart-rending court-led decisions about turning off life-support machines in cases of severe, prolonged coma – Tony Bland injured from collapse of a UK football stadium (unconscious for 10 years) and the young Italian woman Eluana Englaro involved in a road traffic accident (on respiratory pump for 17 years). Without any continuing support, both died rapidly after 3-4 days. At subsequent post-mortem, Bland's brain consisted of an enormous liquefied blood clot together with bits of necrotic brain tissue. Yet the moral uncertainty arising from these (court) decisions, however, has remained somewhat unresolved, and hence seems to lie as a continuing burden on the social conscience.

Although never deployed by others, I think the answer here is to use comparative controls, such as the IRA (Irish Republican Army) hunger striker Bobby Sands imprisoned in the Maze Prison, Belfast, Northern Ireland. He was a fit young man of comparable age to the other two cases yet without any brain injury. But he died ~60 days after commencing his fast: that is the difference. Likewise, the Oxford based UK campaigner for assisted dying Dr Jean Davies took almost 40 days to die although hastened in this case, it should be noted, by her refusal of water during the last two weeks of her life.[18] And she was 68 years-of-age. Clearly unlike Sands and Davies, the other two cases (Bland and Englaro) were now simply resuscitated corpses, since they were both completely brain dead – a crucial difference not widely appreciated by the public, churchmen, or ethicists. Even the judges who authorised the turning off of the switches probably didn't know either. On the grounds that I have offered, there is no reason why their respirator pumps and switches should not have been turned off – maybe even earlier.

In contrast to dense comatose states there are two other relevant conditions, the vegetative state (VS) and the minimally conscious state (MCS). Either commonly arises from external head injuries, internal haemorrhage, or thrombosis. The pathology of VS may be diffuse although the brainstem is capable of preserving heart rate, breathing, and the sleep-wake cycle. Therefore, clinical behaviour subsists in 'awakeness without external environmental awareness', whereas MCS patients, in addition to basal wakefulness like VS, demonstrate variable degrees of alertness to their environment, thus revealing a limited sense of self-awareness in addition to some receptive sensory perceptions from outside.[18]

We can recall the recently reported case of a man thought to be in a chronic vegetative state who had lain in a Belgian hospital for 23 years. He was in fact conscious with 'akinetic mutism', and thus was unable to communicate, even with eye-blink signals. Studied by Dr Steven Laureys (University of Liège, Belgium), who has been performing brain scans on such patients, the true picture was realised. Brain scans have now revealed that many patients thought to be chronically unresponsive are, to varying degrees 'conscious', although not showing any active (motor) response to commands.

From that, two outcomes of importance become apparent. Firstly, that the descriptive 'permanent' in cases of VS is no longer applicable and should not be used. Secondly, that when doctors proclaim someone to be dead, based on very simple bedside criteria typically requiring verbal, reflex, or motor responses to a stimulus or command, *they may well be wrong*. This, for example, arises in cases of cardiac collapse associated with near-death experiences, from which it is uncritically claimed that the phenomenology can only occur with a 'dead' brain.

Now, VS is considered a state of wakefulness only, while MCS may have, in addition, features of self and environmental awareness, even though there may be no speech involved. VS, therefore, is not unlike the developing early foetal brain which first may manifest wakefulness, followed by progressive stages of self and environmental awareness (temporarily akin to MCS) through womb to infancy, thereafter representing growing up and acquiring the fuller attributes of an adult mind.

The two situations (VS and MCS) seem to correlate with other neuroscientific research indicative of two broad levels of cerebral functioning – a resting or 'default' state, and the other representing normal cognitive activity. It is from these positions that we can grasp some meaning of what conscious-awareness and

conscious-activity require in terms of brain circuitry and connectivity, and how these differ from states where consciousness is
either reduced to a variable degree, or indeed non-existent, as
with the varied severe brain injury cases mentioned above.

The Cerebral 'Default' State

It has now been fully established that the quiescent brain, when
not engaged in any specific activity but without being asleep, is
far from inactive. Indeed, this resting activity is a state termed
the 'default mode network' (DMN).[19] While in many contrived
experiments, 'activations' of certain brain areas could be demonstrated with goal-directed activities (as expected), it had also
been noted that in similar circumstances, *decreases* in brain
activity were observed. These decreases were (always) explained
away in various ways, as though unimportant. However, because
of the regular manner in which these 'de-activations' seemed to
occur, it became necessary to recognise that there is a constant
baseline level of neural activity and hence a need to elucidate its
meaning. The DMN, although operative in the circumstances
noted above, is actually a highly metabolic state, consuming 50%
energy supplied to the brain (which, on a daily basis, consumes
20% total bodily calorie intake). This energy is utilised, probably
in the maintenance of synaptic connectivity, and therefore
particular circuitry, throughout the brain. We therefore need to
see what circuitry is involved, and secondly to decipher precisely
what this high-energy DMN system is doing. Biologically, one
might suppose that it is doing important work because although
not alert to incoming cues, it is still highly functional – and at
great metabolic cost.

The highest metabolic activity resides in posterior medial
(inner rear) areas of the cerebral cortex (parietal, temporal and
occipital) employed in visual-spatial processing which orients
the subject both to external environment and internal milieu, in
other words, the natural habitat. This, in evolutionary terms,

makes sense as it means that the brain is continuously monitoring that habitual world for predatory attack, based on rapid perceptions of motion, sound or odour. Other specific parts of this assemblage (posterior cingulate and retro-splenial cortex) are related to emotion and to the emotional valence of past memories relative to the individual.

The posterior lateral (outermost rear) cortex (PLC) has two functions. The right PLC is sensitive to events that seem unfamiliar or unexpected, while the left PLC operates in the recall of memory. Thus together, this area orients the subject to salient alterations, and novel environmental stimuli, related to social cueing and responses to others in the near vicinity. The ventral medial (lower innermost) aspect of prefrontal cortex (VMPFC) surveys sensory aspects of the internal and external environments, and connects with the emotional-functioning limbic system, so coordinating visceral and motor responses with incoming information. The dorsal medial (upper innermost) PFC (DMPFC), while de-activated during goal-directed behaviours, is activated when monitoring one's own mental states, such as self-generated thoughts, planned speech, and emotion. Secondly, this region attributes states of mind to others, thus representing 'Theory of Mind' (TOM) activity. So the DMPFC represents the self – its internal aspects and relations towards others' mental states.[20]

These discoveries, over the last ten years, have opened up new ways of thinking about brain structure and function, beyond the classical descending motor and ascending sensory systems. The data emphasise the extensive connectivity within the brain, bringing together distant and local regions in low-frequency generated systems. At least ten fibre-tract systems have been identified and with refinements, more will be deciphered, associated with specific nodal 'hotspots' throughout the cortex.[21,22] One of the most metabolically active hotspots is located on the inner side of the parietal lobe, termed the precuneus.[23] This lobule is de-activated during sleep and anaesthesia, suggesting that it may be an important contributor to the origins of conscious-awareness.

However, studies with VS patients show the presence of an attenuated DMN, but since these people are incapable of voluntary action (alertness), rather than simply being in a wakeful state, it follows that DMN, and in particular the precuneus, is unlikely to be the source of conscious-awareness.[24,25,26] Conversely, in MCS patients, it has been suggested that restoration of thalamo-cortical connections is apparently associated with return of conscious-awareness (albeit severely impaired).[27]

Computerised scanning of the brains of very early pre-term neonates (~25 weeks gestation), revealed five functional systems, although white matter connectivity (through a technique called 'diffusion tensor' magnetic resonance scanning) showed reduced organisation compared with adults.[28] Foetal thalamo-cortical connectivity is said to originate around 25 weeks gestation, so it is clear that if these infants were scanned regularly as they come to term, it should be possible to define their brain growth, and hence infer what capacities were coming on-line, relevant to age. But whether consciousness (that is, primary awakening) is related to ascending impulses from the upper midbrain ('reticular formation') is not answered by these studies, neither whether it is the precuneus that plays a crucial role in this outcome. Given the importance of the upper midbrain and pons in providing essential ascending impulses to the thalamus and cortex beyond, its role should not be forgotten.[29] However, other studies have shown that consciousness can be suspended by electrical stimulation of a cortical area deep within the infolding between the frontal and parietal lobes. This is known as the insula, and beneath that, a thin strip of neural tissue called the claustrum, which seems to influence significantly levels of consciousness.[30,31] These are exciting discoveries embodying techniques which may well help to delineate the fundamental neuro-structural and neurophysiological criteria for the origins and operations of awakeness and alertness.

Grappling with Being Conscious

The notion of being conscious is critically emphatic of the first person perspective. Within that private, first person subjective sphere of being conscious, different states of awareness are recognised – thought, internal speech, self-reflective introspection, imagination and self-consciousness.

Yet one of the fascinations of our day-to-day conscious life is that it is so illusory. Throughout much of our lives, many processes are effected without input from conscious-awareness, such as dressing oneself, driving a vehicle in dense traffic, or even, as pianists tell us, playing a concerto from memory in a large symphony hall. Fully directed attention only intrudes with an irregularity – a missing shirt button, or flashing blue lights on the carriageway, or a sudden thought that some task needs to be undertaken. The more skilled we become in performance, the better we act on a subconscious plane. Furthermore, many fundamental metabolic processes are outwith conscious control such as respiration, heart rate, blood pressure, hormone secretion, fluid balance, or digestion.

Despite that, we do not think life is a film show or a dream. Conscious-awareness is a reality even though that reality is dependent on the brain. For example, if we think of the continuous, rapid scanning sweeps of our eyes – or saccades, it is estimated that about twenty per cent of our visual life is 'blind'. Again, if we think of blinking and the rapid bobbing about of our eyes, there is no direct relationship to what we thought we saw, or are seeing: instead we perceive a steady, smoothed and apparently continuous play-back which is a cerebrally and not a retinally engineered event. We see with our brains, and not our eyes. Moreover, there is no 'picture' of any exterior panorama within the brain. Our view of life is effected through a subjective port which we sense is located behind our noses and at the point where our heads turn laterally. So we can look out but not see our faces (other than through means of a digitised 'selfie' – I think that is the right word), or our backs, or anything internally. Our virtual body-image, contrived from visual, vestibular and joint and skin inputs, is welded into a sense of relationship with the environment, and of the environment to us, by the lower aspects of the parietal lobe.

Physiologically, our brains can be deceived, and there are various pictures that either mean one thing, or another. Electric currents of a certain frequency applied to muscle tendons make us feel that the muscle is moving and doing things that are anatomically impossible (the 'Pinocchio' effect). Pathological events in the brain upset our sense of body image. Migraine sufferers may sense a duality of presence – *there am I and here am I*.[32] Similar artefacts of personhood accompany the auras of temporal lobe epilepsy.

Most people are familiar with the Phantom Limb phenomenon – the perceived presence of a part which is no longer anatomically present – such as a leg, hand, or breast: in other words – the neurophysiology of absence. But this also happens with children born without certain limbs – one child when she was old enough to give a verbal report said she used her phantom fingers for doing her sums with. This, however, implies that we inherit a generalised engram of our body image which has to be constantly updated and modified throughout life, since our bodies continually change shape.[33]

But can we ever explain the origins of a meta-physical entity like conscious-awareness from its brute anatomical, electro-chemical, and frequency-driven oscillatory networks? There are a lot of neuroscientists who think so and are putting all their efforts into solving the problem. If there is any merit in Daniel Dennett's book *Consciousness Explained*, it is his intent to excise the Cartesian duality out of consciousness thinking – the Rylean 'ghost in the machine'. But theories abound. We have parallel multiple drafts circuitry (Dennett and Kinsbourne), re-entrant circuits (Edelman), oscillatory circuits at 40+ Hertz (Llinas and Steriade), searchlights (Crick and Koch), and more recently the global workshop (Baars) – which seems the most likely theory to survive. Then we have quantum wizardry operating around neuronal microtubules. But microtubules are not the exclusive preserve of neurones – so why do we not see consciousness arising out of the kidneys or intestines, whose cells also contain microtubules?

One might politely ask whether these are the only ways of tackling the problem. Large systems functioning in harmonic unity do not acquire consciousness: for example, the (UK) National Grid is a vast array of parallel-distributed cables which oscillates at 50Hz: yet this vast system of wires is not thought to have achieved conscious-awareness. How is one of many parallel circuits, or multiple drafts of Dennett, selected for entry, or re-entry – as in Edelman's model – into consciousness, and who or what does the selecting? Or functions – like Crick and Koch's searchlight? How does the searchlight, based on the reticular nucleus of the thalamus, incorporate olfaction (smell), which is an important contributor to phenomenal life and yet bypasses this important co-ordinating centre?

The Meanings & Value of Conscious-Awareness

From what I have tried to describe, it is clear that a precise definition of what consciousness is, and how it relates to the brain, is no easy task. It seems likely that consciousness arises during foetal development, but its temporal or structural details are not known with any certainty. I have stressed the foetal model because, heuristically, I think it has much to contribute to our understandings of what is entailed in becoming conscious, as well as to our being conscious.

Much neurophysiological research involves defining functional anatomical locations invariably within the cerebral cortices, an approach which continues to facilitate the view that there are 'neural correlates' of consciousness, and that these correlates – necessarily again – it seems, have to reside somewhere in the cerebral cortex, while the universal role of the upper brainstem in modulating much of this activity seems to be completely forgotten.

However, as often seen in science, approaches from completely different vistas, and seen from different thought-ideas, often bring new heuristic insights to a pre-existing 'hard-nut' problem. Here I am thinking about those who use robots modelled on insect neurological systems. In these new models, the robotic electronic 'brain' does not know its environment – and neither does the human brain. They both have to discover facts about that environment by relating movement to incoming information received through external sensors. Most importantly, the mathematical portrayal of these systems reveals how a large number of unwieldy pieces of information can be reduced to a few principles such as to how the 'organism' works – and, importantly – learns. This type of innovative modelling is not far distant from the foetal brain in its preliminary steps, learning its environment (and more!) – a process which, as we have seen, takes up to one-quarter of the life-span of an adult to have been accomplished effectively.

I am also attracted, for example, by the work of O'Regan and Noë in emphasising that the brain does not create an internal model of the subject's environment.[34] They resist attempts to define the so-called 'neural correlates of consciousness', whether considered as multiple parallel pathways, re-entrant phenomena, frequency-bound systems, or quantum activities within microtubules. For them, to be consciously aware is to be on the move: we move to think, explore and hence to know our environment, and upon which our ability to cope with it, and to understand it, are of crucial importance. And if you do not think that is true, just watch a three-month-old baby at play. We have to know fully the accompanying sensori-motor features about spheres, plates and faces in order to perceive them for what they are. And much of those simpler axioms are learnt as a result of mobile play from our earliest moments from birth onwards – through time and beyond. A mind requires a brain that is embodied in a moving person, who is further embodied in a spatially-moving environment. Much that is written about conscious-awareness, or of being conscious, fails to incorporate that highly relevant motor aspect of our behaviour.

Beyond those abilities acquired through the appropriate sensori-motor contingencies of life, come the more abstract features of thought, music, mathematics, art. This raises the question as to why we need so much intelligence! To paraphrase Professor Polkinghorne (Cambridge University) – humans are smart enough to interrogate the universe and elucidate its lawful workings. But why do we need to be able to do that? It has no immediate evolutionary benefit, nor value for our day-to-day biological existence. How many of us really understand the Big Bang, how stars are formed, or anything worthwhile about the murky, insubstantial world of quantum mechanics? And apart from myself, how many people have actually read Stephen Hawking's book *A Brief History Of Time*? Very few of us, I should venture. And if we did know, would it have much personal relevance to, and for, our lives and relationships?

Raymond Tallis observes that conscious being is at the very heart of personhood, underpinning its immediate responses to all the contingent and unexpected challenges of a constantly moving and changing environment.[17] It is this notion about the sheer centrality of conscious-awareness – indissolubly welded to the person at its centre – about which Tallis so excitedly, and correctly, exults. This sense of person-as-the-centre-of-consciousness, or as I said at the beginning, of *conscious-beings being conscious* contrasts markedly with the existing models which construe consciousness as an appendage of being – something 'in-the-brain' or 'out-of-the-brain'.

Seen from a wider perspective, there is no 'progress' in evolution because evolution is blind and knows no progression. Despite that, we see incremental enlargements in our genome, in bodily size, and in our vast brains. Yet it *is* that gradual process, through evolutionary time, which has conferred on us our massive cognitive abilities within the scope of being conscious and acquiring conscious-awareness. And it is the acquisition of conscious-awareness that has paradoxically empowered us, in our highest cognitive modes of thought, to use our brains (apparently) to discern new knowledge and ideas, to ponder our presence in the universe, and to know that we shall die.

So, has that process been entirely blind and without purpose, or can there be discerned within it another purpose brought about by additional, unseen influences contributing to this incredible story? That is what we have to ponder in due course (Chapter 6).

The late Kathleen Wilkes, philosopher of St Hilda's College, Oxford University, loved to tease those who were concerned so much about consciousness.[35] In her opinion, the use of the word conscious (being so meaningless) should be dispensed with. However, in one of her papers, she offered a quote from another source (a source I am unable to trace): 'Consciousness is like the Trinity: if it is explained such that you understand it, it hasn't been explained properly'.

Here, then, I should also hope that my account of consciousness has not been sufficiently articulate, thus to secure the reader's complete understanding.

But, at least, we are beginning to know quite a lot about it.

* * *

1. Zizioulas, J, 1985. *Being as Communion*. New York: S Vladimir's Seminary Press, 39.

2. Chalmers D, 1997. *The Conscious Mind*. Oxford: Oxford University Press 1997, 293.

3. Boycott B, Young J, 1955. A memory system in Octopus vulgaris Lamarck. *Proceedings of the Royal Society of London B* 143: 449-480.

4. Griffin DR & Speck GB, 2004. New evidence of animal cognition. *Animal Cognition* 7: 5-18, 2004.

5. Seth AK, Baars BJ, Edelman DB, 2005. Criteria for consciousness in humans and other mammals. *Consciousness and Cognition* 14: 119-139.

6. Kieffer J, Colgan P, 1992. The role of learning in fish behaviour. *Reviews in Fish Biology & Fisheries* 2: 125-143.

7. Tate A, Fischer H, Leigh A, Kendrick K, 2006. Behavioural and neurophysiological evidence for face identity and face emotion processing in animals. *Philosophical Transactions of the Royal Society B* 361: 2155-2172.

8. Kendrick K, da Costa A, Leigh A, Hinton M, Peirce J, 2001. Sheep don't forget a face. *Nature* 414: 165-166.

9. Cavalieri P, Singer P (eds), 1993. *The Great Ape Project*. New York: St. Martin's Press.

10. Northcutt RG, 1996. The Agnathan Ark: the origin of craniates brains. *Brain, Behaviour and Evolution* 48: 237-247.

11. Cabanac M, Cabanac AJ, Parent A, 2009.The emergence of consciousness in phylogeny. *Behavioural Brain Research* 198: 267-272, 2009.

12. Heyes C, 1993. Anecdotes, training, trapping and triangulating: do animals attribute mental states? *Animal Behaviour* 46: 177-188.

13. Heyes C, 1994. Reflections on self-recognition in primates. *Animal Behaviour* 47: 909-919.

14. Hepper P, Shahidullah S, 1994. The beginnings of mind –

evidence from the behaviour of the fetus. *Journal of Reproductive and Infant Psychology* 12: 143-154.

15. DeCasper A, Fifer W, 1980. Of human bonding: newborns prefer their mothers' voices. *Science* 208: 1174-1176.

16. Mampe B, Friederici A, Christophe A, Wermke K, 2009. Newborns' cry melody is shaped by their native language. *Current Biology* 19:1994-1997.

17. Tallis R, 1999. *The Explicit Animal: A Defence of Human Consciousness.* Basingstoke: Macmillan Press.

18. Giacino J, Ashwal S, Childs N, et al, 2002. The minimally conscious state: definition and diagnostic criteria. *Neurology* 58: 349-353.

19. Gusnard D, Raichle M, 2001. Searching for a baseline: functional imaging and the resting human brain. *Nature Reviews Neuroscience* 2: 685-694.

20. Frith C, Frith U, 1999. Interacting minds – a biological basis. *Science* 266: 1692-1695.

21. Damosieaux J, Rombouts S, Barkof F, et al, 2006. Consistent resting-state networks across healthy subjects. *PNAS [USA]* 103: 13848-13853.

22. Buckner R, Sepulcre J, Talukdar T, et al, 2009. Cortical hubs revealed by intrinsic functional connectivity: mapping, assessment of stability, and relation to Alzheimer's disease. *The Journal of Neuroscience* 29: 1860-1873.

23. Frasson P, Marrelec G, 2008. The precuneus/posterior cingulate cortex plays a pivotal role in the default mode network: evidence from a partial correlation network. *Neuroimage* 42: 1178-1184.

24. Boly M, Phillips C, Tshibanda L, et al, 2008. Intrinsic brain activity in altered states of consciousness: How conscious is the default mode of the brain? *Annals of the New York Academy of Sciences* 1129: 119-129.

25. Boly M, Tshibanda L, Vanhaudenhuyse A, et al, 2009. Functional connectivity in the default network during

resting state is preserved in a vegetative but not a brain dead patient. *Human Brain Mapping* 30: 2393-2400.

26. Vanhaudenhuyse A, Noirhomme Q, Tshibanda L, et al, 2010. Default network connectivity reflects the level of consciousness in non-communicative brain-damaged patients. *Brain* 133: 161-171.

27. Laureys S, Faymonville M, Luxen A, et al, 2000. Restoration of thalamocortical connectivity after recovery from persistent vegetative state. *Lancet* 355: 1790-1791.

28. Fransson P, Skiold B, Horsch S et al, 2007. Resting-state networks in the infant brain. *PNAS [USA]* 104: 15531-15536.

29. Urbano F, D'Onofrio S, Luster B, et al, 2014. Pedunculopontine nucleus gamma band activity – preconscious awareness, waking, and REM sleep. *Frontiers in Neurology* 5:1-12.

30. Koubeissi M, Bartolomei F, Beltagy A, Picard F, 2014. Electrical stimulation of a small brain area reversibly disrupts consciousness. *Epilepsy & Behavior* 37: 32-35.

31. Kurth F, Zilles K, Fox P, Laird A, Eickhoff S, 2010. A link between the systems: functional differentiation and integration within the human insula revealed by meta-analysis. *Brain Structure and Function* 214: 519-534.

32. Marsh MN, 2015. Hey! What's that gorilla doing over there? On the Illusory-Hallucinatory Nature of Everyday Living. *European Review*, in press.

33. Brugger P, Kollias S, Muri R, et al, 2000. Beyond remembering: phantom sensations of congenitally-absent limbs. *PNAS [USA]* 97: 6167-6172.

34. O'Regan JK, Noë A, 2001. A sensorimotor account of vision and visual consciousness. *Behavioural and Brain Sciences* 24: 9390-1031.

35. Wilkes K, 1984. Is consciousness important? *Brit Journal for the Philosophy of Science* 35: 223-243.

Chapter 5

On Acquiring Language

One of the most distinctive features of human beings is their capacity for acquiring language, and the communicative outcomes resulting from that predisposition. Furthermore, it is likely that our ability to engage in abstract thinking is predominantly based on the ability to speak and employ language. Animals do not have the ability to speak, or even to be taught to speak, as amply demonstrated by several attempts with various target animals. But they are able to make specific calls to warn, identify, or warn off aggressors. That evolutionary capacity has to be noted in order to understand the origins of our own verbal capacities. The nature of inter-communication between whales is not understood, while talking parrots are exceptionally good at hearing sounds and repeating them – often to the embarrassment of their owners.

We do not have to be expert linguists to grasp the almost unbelievable facility of babies and young infants to acquire speech, such that by their third year, they can construct meaningful sentences, supported by a reasonable vocabulary. In the previous chapter, we saw that the mother's voice and the prosody (rhythm and accentuation) of her native language are perceived by the later-term foetus, thereby ensuring continuity during the earliest phases of life beyond the womb with a mother only heard but never previously seen. One of the instinctive aspects of the way in which humans vocalise with very young babies is to employ 'motherese', with higher pitched frequencies, longer vowels, softer delivery, and a slower rate of speaking. Also with that comes another instinctive attribute – that of pausing for the baby to make a facial, or verbal response whatever the form of 'babble' emitted – thereby initiating and ensuring a kind of you-and-I conversational style that is established from the very beginning.

Although when we think about language, the emphasis is towards talking and speech, it must be remembered that in order for those abilities to develop, it is necessary to hear the characteristic sounds of the language to be learned, and that requires aural capabilities which can monitor the relevant sounds, and employ them in the speech learning process. However, deaf babies can master sign language, while deaf children not previously introduced to sign language from an earlier age, still retain the ability to construct their own form of signing. That is not much different, one might suppose, from our attempts to order a cup of coffee or find out what time a certain train leaves the capital, when one cannot speak the local (foreign) language efficiently. Nevertheless, the astounding ability of very young humans to develop speech without formal education or much conscious effort is a remarkable phenomenon. On the other hand, the understanding of the biological processes whereby that facility is gained has been shown to be an enormously difficult, and far less tractable, a problem.

Indeed, over decades, linguists have analysed speech into its basic sounds (phonemes) and the way in which they are put together informing more complex words, their conjunction into meaningful sentences (morphology) and how those constructions are governed by grammatical rules (syntax) and the meanings (semantics) thereby conveyed to the listener. On the other hand, physiologists have tended to focus on the two major, classical regions of the brain thought to be centrally involved in speech acquisition, Broca's Area (inferior frontal gyrus) and Wernicke's Area, occupying the superior temporal gyrus, and their interconnecting white matter (or arcuate fasciculus), although that is no longer sufficient. But in order to understand the broader sweep of language and how it has come to us as human beings, far more attention needs to be paid to its earliest historical, evolutionary roots – or the descent from progressive modifications, as suggested by Darwin [1] In recent years, that approach has been increasingly amplified by advances in brain scanning, molecular genetics, and comparative archaeology. As a result, we have a more detailed understanding of why humans have language and can speak, although the picture is by no means complete.

Language & Speech

There is a subtle difference between language and speech. Language might be generally conceived as a cognitively-based construct, contributed to, and represented by, the laws of grammar, the extant vocabulary (lexicon), and its finer usage in eloquent literature and poetry, and hence demonstrative of the cultural attributes of a society, people, or race. Another aspect of language is that it can be written down, which presupposes a corresponding alphabet and words which can be stored as a lasting document, and read whenever convenient, another facet of language acquisition which has to be appropriated as an abstract discipline. Thus conceived, language is akin to musical notation, composition and performance, such that the notes (pitch and length) are analogous to phonemes, measured phrases to sentences, and time signatures comparable to the speed of words delivered to an audience. Mathematics could be conceived as an even more abstract form of dealing with notional quantities and their lawful relationships. Thus, it seems more reasonable to regard speech elements as gestures, accompanied by appropriate facial and bodily indications which both serve the dual tasks of (motor) production, and (sensory) perception.[2]

Speech and talking, on the other hand, represent the basic, day-to-day communication skills necessary for the essential commerce between individuals and organisations. In other words, speech, to a varying extent, conveys language. But, as noted above, if you are unfamiliar with the local tongue, communication can be rather difficult, while infants can pick it up through constantly hearing and recognising the correct sequence and hence meaning through stress recognition, and by distinguishing between the subtle differences in length of sounds.

The evolution of speech is obviously anterior to the conceptualised domain of language, with the need for the anatomical processes of articulation, the parallel ability to hear the sound frequencies emitted by means of an adapted auditory system, and the neurological correlates capable of co-ordinating sensory (sound) inputs, motor outputs, and their cross-representations both across dedicated centres of the brain, as well as into conscious-awareness.[3] Obviously, such a complex system requires genetic control, yet when viewed in evolutionary terms, its molecular basis is, so far, poorly understood. Moreover, since speech does not become fossilised, the evolutionary history of speech must therefore rest on comparative anatomical details observed not only between animals and humans, but also between each successive layer of humanins, as can best be ascertained from the extant archaeological record.

Of major importance has been the recognition of the anatomy of the oropharyngeal tract and its production of 'formants', or resonances.[4] While all mammals employ the same laryngeal system, only humans use formants as the underlying production basis of communicative speech. This is exemplified, par excellence, in producing a pitchless whisper, which does require laryngeal vibrations, yet is still intelligible, similar to computerised synthetic speech. Whispering is exclusive to humans, and is a particularly difficult task for actors to acquire and execute on stage. Our ability to produce so many formants rests with the very low position of the human larynx, and with the fact that human vocal tract anatomy is dissimilar to that of other closely related primates.[5] Because of differing facial anatomies, primate mouths are long, while in humans the mouth is relatively short because our jaws do not protrude, while the trachea and larynx are elongated in the vertical dimension. Thus, comparatively, the available 'vowel space' of nonhuman primates is extremely restricted.

Furthermore, the tongue projects downwards towards the upper reaches of the larynx, allowing the creation of vowels such as 'i' as in beet, and 'u' as in boot. The descent of the larynx, while an ancient trait, only occurs phenotypically at about three months, thereby allowing the baby before that time to suckle and breathe simultaneously. It is this changed position of the larynx which allows humans the wide repertoire of formants employed in speech, and is thus, perhaps, the key event in speech evolution.[5]

The second most important advance was the greatly enhanced ability in employing tongue, lips, teeth, palate (in addition to the vibrations of the vocal cords) to compose the many sounds (phonemes) of all human languages. This exquisite neural control permits, for example, the articulation of the letters 'p' and 'b', the latter subtly longer – but by only ten milliseconds – yet this difference, amazingly, has been demonstrated experimentally to be perceived by the ears of very young babies.

A third key element fundamental to speech acquisition is that of imitation. The specific anatomical changes outlined above endow humans with the power to imitate widely, not just words, but also musical sounds, and to whistle. Such outcomes (apart from a few taught changes) are structurally impossible with animals – domesticated or in laboratories. So, as animal culture does not change and neither does their language, they lack the innovative changes which characterise human articulatory prowess and literary expression, and the resultant cultural and technological advances. There are some notable exceptions, especially with the learning of bird song, which is an important part of territorial governance, and as means of attracting potential female mates.[6] Apart from some specific examples, however, imitation is unusual among primates.

This brings us to the observation that neural commitment to speech is critical in early learners, related to a 'sensitive' period during which language acquisition is most optimal. This includes *categorical perception*, the ability present from birth which enables distinctions between different vowel sounds and the subtle changes in length of varied sounds to be discriminated. Moreover, categorisation occurs despite marked differences in pitch (contrast between male or female speech), speed, articulatory clarity, even 'accent'. But within a year, infants can no longer recognise sounds other than those characteristic of their mother tongue: this does seem to be a unique feature of humans compared with other primates.[7] By 18 months, infants can understand around 150 words, and can vocalise about 50.

But how they distinguish words which are always run together in speaking, is not clear and has not been entirely elucidated. Adults encounter similar difficulties when listening to a foreign language – it seems to be spoken too quickly, unlike the reading of words that are separated from each other.

Consider: THEREDONATEAKETTLEOFTENCHIPS.

Could those connected letters mean: 'the *red* on a *tea ket*tle *of*ten *chips*' or '*the*re, do*nate* a *ket*tle of *ten chips*'?.[7]

For speech learning, attention has to paid, in part, to accentuation, since in English, the accent tends to be on the first syllable. The other important feature is that of social influence and the interactions especially between mother and child. If that crucial interplay is lacking from the beginning, no speech is acquired, and later attempts to correct the deficiency do not entirely succeed. And there, as we have already seen, imitation is seen as an additional, crucial requirement. That has been exemplified with the case of Genie, the Los Angeles girl locked up in solitary confinement for nearly 15 years and hence bereft of those necessary early ingredients for language acquisition to take place.[8] Although rescued and taken to hospital, her speech never developed normally, despite her learning some words and being able to string them together in some kind of intelligible manner. This is the case for all feral children ultimately brought back into civilisation.[9] Genie reminds me of Karzi, the chimpanzee whose linguistic accomplishments, after intense training activities, were not far different.

In order to progress further from this position, we now need to explore the historical record for evidence of how speech began: our observations take us further back than the conventionally quoted 50-100 thousand years and taken to indicate a sudden emergence of speech during the Upper Paleolithic epoch.[10,11]

Excavating the Fossil Record – When Did Speech Begin?
There are, obviously, no remains of speech, while written words were not in vogue during the periods that form the material of this investigation. In attempting to make sense of human speech evolution, a comparative perspective is needed, both with earlier hominins as well as with other evolving primates. In an extensive review, it was argued that *Homo heidelbergensis* was well established by 500,000 years ago, with a sophisticated inventory of tools, javelins, killing of large game, and use of pigments. From them, it is likely that *Homo neanderthalis*, *Homo denisovan*, and *Homo sapiens* arose, and co-existed over a long period of time, as well as inter-bred, each having 23 chromosomes in their genome, sharing 91% human accelerated regions (HARs) and, importantly, having the same pair of specific mutations (associated with speech) in their FOXP2 gene.[13,14,15]

More importantly, the Neanderthal pelvis is of modern shape, and could accommodate a modern sized baby of equivalent brain size.[16,17] Moreover, it is likely that these hominins lived in patrilocal small groups with a birth rate every three years. Therefore, Neanderthal life was of a similar tempo that we, as modern humans, would be accustomed to, with bi-parental care of children, some semblance of a midwifery service, and growing up on a similar time frame, as indicated by developmental dental eruptions.[18,19]

In regard to speech perception, reconstructions of the

anatomy of the external and middle ear cavities of early human species showed their maximal frequency responses to around 4kHz (range 1-6kHz), as typical for humans, although the fossils in question derive, significantly, from *Homo heidelbergensis* (Sima de los Huesos, Spain), while the ossicles (middle ear) of a group of Neanderthals did not differ significantly from those of modern humans.[20,21] These findings indicate that the anatomy of perception as far back as the Heidelbergs (>500,000 years) was little different from modern humans. In terms of sound production, the larynx had already descended, the hyoid bone was now solid, and the spinal cord was enlarged in the region of the thoracic cage, indicative of considerable use in controlling breathing which (although suggested by others might have subserved an aquatic use) is in keeping, one might have thought, with a necessary regulation of subtle respiratory efforts in the acquisition, maintenance, and control of speech.

With this brief resume of the relevant facts covering enormous fields of anthropological research, it seems very plausible to suppose that language began its evolutionary ascent very early in hominoid history, most likely involving *Homo heidelbergensis* more than half a million years ago. This seems to make more sense, viewed in evolutionary terms, than to suggest that speech was a sudden, out-of-the-sky event that suddenly intervened in recent human affairs about 50,000 years ago. Furthermore, one has to take into account the incomplete developments in animal articulation, which, although not finding another convergent outlet for speech would, undoubtedly, have contributed significantly to that outcome for humankind.

Yet in addition, the most important and overriding genetic basis for that claim rests on molecular analyses of the FOXP2 gene. The discovery of FOXP2 and its relationship to speech was a serendipitous occasion, brought to light by discovery of a three-generation inbred kindred, of whom about 50% (in strict Mendelian dominant, monogenetic fashion) exhibited a profound articulatory deficit in language production.[1] FOXP2 is one of the most highly conserved (non-mutation prone) genes known (in the top 5%) and has remained unchanged over 100,000 years from rodents to the recent split between chimpanzees and humans from their last common ancestor, 6.5 million years ago. Yet, during that recent eye-blink period seen in terms of evolutionary time, the gene has undergone mutational changes at two loci thus directing a change in amino acid composition (T303N, N325S). However, since these changes are present in Neanderthal genomes, these mutations occurred, at least, before 400,000 years ago.[22,23]

This is not to suggest that FOXP2 is the long-awaited 'gene' for speech, but rather to indicate the historical evolutionary origins of speech and with it, the development of widely-dispersed cerebral pathways underwriting this capacity, the 'truncated' convergence of some of these systems in other parts of the vertebrate domain with songbirds, echo-location in bats, and dolphin communication, and its co-ordination with many other neuromuscular and sensory functions which, in part, have led to the origins of human speech/language – and of what we know and (so little) understand of it today.

From that background, there is no evidence to indicate that human speech is a unique phenomenon, since genes, items of the neural circuitry within the brain already existent in earlier species, including those precursors of modern man, have played their roles. However, it might be concluded that the evolved perfection of human speech has, in part, permitted the unique cultural and technological advances which characterise human flourishing and which still permit its outpourings.

* * *

1.   Fisher S, Marcus G, 2005. The eloquent ape: genes, brains and the evolution of language. *Nature Reviews Genetics* 7: 9-20.

2. Liberman A, Whalen D, 2000. On the relation of speech to language. *Trends in Cognitive Sciences* 4: 187-195.

3. Fitch W, 2000. The evolution of speech: a comparative review. *Trends in Cognitive Sciences* 4: 258-267.

4. Stevens K, House A, 1995. Development of a quantitative description of vowel articulation. *J Acoust Soc Am* 27: 484-493.

5. Lieberman P 1969. Vocal tract limitations on the vowel repertoires of rhesus monkeys and other nonhuman primates. *Science* 164: 1185-1187.

6. Marler P, 1991. Song learning behaviour: the interface with neuroethology. *Trends in Neurosciences* 14: 199-206.

7. Kuhl P, 2004. Early language acquisition: cracking the speech code. *Nature Reviews Neuroscience* 5: 831-843.

8. Curtiss S, Fromkin V, Krashen S, et al, 1974. The linguistic development of Genie. *Language* 50: 528-554.

9. Newton M, 2002. *Savage Girls and Wild Boys.* London: Faber & Faber, 119.

10. Mithen S, 2005. *The Singing Neanderthals: The Origins of Music, Language, Mind and Body.* London: Weidenfeld & Nicolson, 246.

11. Berwick R, Friederici A, Chomsky N, Bolhuis J, 2013. Evolution, brain, and the nature of language. *Trends in Cognitive Sciences* 17: 89-98.

12. Dediu D, Levinson S, 2013. On the antiquity of language: the reinterpretation of Neanderthal linguistic capacities and its consequences. *Front Psychol* 4: 1-17.

13. Meyer M, Kircher M, Gansuage M, et al, 2012. A high-coverage genome sequence from an archaic denisovan individual. *Science* 338: 222-226.

14. Green R, Krause J, Briggs A, et al, 2010. A draft sequence of the Neanderthal genome. *Science* 328: 710-722.

15. Krause J, Lalueza-Fox C, orlando L, et al, The derived FOXP2 variant of modern humans is shared with Neanderthals. *Curr Biol* 17: 1908-1912.

16. Weaver T, Hublin J-J, 2009. Neanderthal birth canal shape and the evolution of human childbirth. *PNAS [USA]* 106: 8151-8156.

17. de Leon M, Golovanova L, Doronichev V, et al, 2008. Neanderthal brain size at birth provides insights into the evolution of human life history. *PNAS [USA]* 105: 13764-13768.

18. Lalueza-Fox C, Rosas A, Estalrrich A, et al, 2010. Genetic evidence for patrilocal mating behaviour among Neanderthal groups. *PNAS [USA]* 108: 250-253.

19. de Castro J, Rosas A, Carbonell E, et al, 1999. A modern human pattern of dental development in lower Pleistocene hominids from Atapuerca-TD6 (Spain). *PNAS [USA]* 96: 4210-4213.

20. Martinez I, Quam R, Rosa M et al, 2008. Auditory capacities of human fossils: a new approach to the origin of speech. *J Acoust Soc Am* 123: 3606 (doi:10.1121/1.2934784).

21. Quam R, Rak Y, 2008. Auditory ossicles from south-western Asian Mousterian sites. *Journal of Human Evolution* 54: 414-433.

22. Ayub Q, Yngvadottir B, Chen Y, et al, 2013. FOXP2 targets show evidence of positive selection in European populations. *Am J Hum Genet* 92: 696-706.

23. Fisher S, Scharff C, 2009. FOXP2 as a molecular window into speech and language. *Trends Genet* 25: 166-177.

24. White S, Fisher S, Geschwind D, et al, 2008. Singing mice, songbirds, and more: models for FOXP2 function and dysfunction in human speech and language. *J Neurosci* 26: 10376-10379.

Part II

Dignity

Chapter 6

Theological Approaches to Personhood

In the previous chapters, I have sketched out the physical and evolutionary aspects of human anthropology, including analyses of its most important aspects – genes, consciousness, and language. Does all that define mankind, thus adding up to a perspective that is undoubtedly, and decidedly, unique? My answer would be a cautious 'no', in that evolution inevitably means the juggling of ancient genes into something novel, or distinctive, while various degrees of conscious-awareness and a semblance of language – viewed as communication – are neither 'unique' to humans alone. So anything 'new' about human beings is rather more usefully defined as 'distinctive'. Any uniqueness could only be applied to humanity's cultural characteristics which far exceed those capabilities of any other creature within the known universe. But culture, per se, is necessary but not entirely sufficient in defining humankind, or what it means to be a human person. If we press still further, we could ask whether that culturally defined uniqueness is itself sufficient in framing mankind, without remainder. For example, civilisations come and go.

In 2012, the scientific journal *Evolutionary Anthropology* published a well-received mini-symposium under the heading 'What Makes Us Human?'.[1] A perusal of the comments made by its contributory authors reveals coverage of the ground I have already explored in the preceding chapters. We are not genetically 'unique'; but unlike chimpanzees, humans exhibit symbolic behaviour, the ability to plan for the future, have language and a diverse culture which is transmissible to others and thus handed on to children; and exhibit a 'Theory of Mind' – knowing what others believe and think, and knowing when others harbour false beliefs. That cognitive ability is determined by possession of large brains, which facilitate moral decision making, emotional valences appropriate to circumstance, a sense of body image, and the ability to co-ordinate emotional outcomes with incoming afferent bodily sensations. Breeding is much slower, leading to parental care, monogamy, and the phenomenon of grandmothering, since females live beyond their reproductive years and probably contribute to the development of a kind of midwifery and help with the bringing up of offspring. The sociality of humans, fostered through language, reminiscence, and the use of fire as a centre for that sociality, provides a wide-ranging curriculum which points to the uniqueness of humans as a cultural species.

Yet we must not be taken in by that summary of the principal aspects of developed human culture. The reason is that the comments made above are generalisations. There is something vitally amiss with that definition of humanity. It lacks specific reference to the individual, and definitions of what those human individuals are like and how they should be. That I believe to be a requirement of a religious perspective, a perspective related to the divine which is more specific and defined than the usual references to 'the gods' in conventional archaeological parlance.

We need a properly developed theological anthropology of humankind.

Biblical Accounts of Human Anthropology – Hebrew Bible

The Jewish background to person is of a unitary body and mind, and its abrupt cessation at death. In the 'P' (c.600-500 BCE) version of Genesis humankind is created in God's *image* (...bezelem), both male and female, to live as His viceregents on earth, in harmony with each other in an exclusive relationship with Him (Gen 1:26).[2] In the older 'J' (Yahwist) tradition (dating from c. 1,000 BCE), the breath of life (nephesh-hayah) vitalises the earthen vessel destined for humanity (Gen 2:7) thereby establishing both a physical origin, and an end (met-hayah) as God's breath (ruach) departs the corpse (Gen 35: 18). There is no perception of immortality. The hovering in the 'lower pit' (sheol) of weakened bodily 'shades' (rephaim) from which the breath (nephesh) had withdrawn, was a vague concept never elaborated further in the pre-exilic period. A future existence was never envisaged.

The ontological anthropology here is of being resolutely tied to that of a fulsome life lived to old age (Ps.128). This is the prayer of a peasant people: to work the inherited lands of one's father, to have many sons at one's side, to see the fruit of one's toil and to be fulfilled in the abundance of one's reaping and gathering.[3] Any sense of continuity emerged through embodiment of one's name, through one's sons, in perpetuity, thus to contribute to the collective prosperity of Jerusalem.

Interestingly, there was no definitive word for a body in biblical Hebrew: over eighty body parts are described each of which, by synecdoche, is capable of alluding to the entire person.[4,5] Here, then, 'man is conceived, not in some analytical fashion as a 'soul' and 'body' (flesh or basar), but synthetically as a psychophysical whole'.[6] Furthermore, the person is never isolated, individualised or regarded as a thing, but lives in continuity with others – the so-called corporate nature, or corporate personality, of Israel.[7]

In moving into First-Century Palestine and the Diaspora, we enter the milieu of Hellenistic culture and philosophy and its interactions with Jewish thought-forms and the new Christian thinking. Are new vistas encountered in giving account of Pauline and other New Testament anthropologies? David Stacey, for example, notes the large number of Greek words adopted into contemporary Hebrew writings, including thirty in the Misnah: even Josephus was forced to learn Greek in order to keep up with commerce and officialdom in this widening environment.[8]

In entering this new territorial domain, one major interpretative difficulty now entails critical adjudications on meanings of important word usages such as soma, sarx, psuche, and pneuma, and their construal with Hebrew concepts embodied in words already met, such as nephesh, ruach, and basar, particularly within the Pauline corpus and later gospel accounts.

Biblical Accounts of Human Anthropology – New Testament

First, the Pauline view of mankind. This is not the easiest to discern from a 21st-Century perspective. Although Paul was a staunch Hebrew and subject to a strict upbringing in the synagogue, he was raised in a Greek-speaking milieu. Greek would have been his first tongue, although Christianity rudely intruded into his life to interrupt it forever. In his writings the use of Greek words are not to be necessarily construed in their native meanings, but in the special understandings which he attributed to them in enabling him to disseminate the gospel. Yet his outlook was neither analytic nor 'partative', but conclusively and consistently emphatic of the unitary nature of mankind, thus reflecting the underlying Semitic cast of his mind.[9,10] As a Jew, and a proud Benjamite who had excelled in his studies, Paul's thinking was centred on God, not on man *qua* man in reflecting the prevailing Hellenistic temperament.[11] Having discovered a new view of God on the Damascus road, his subsequent task was to articulate that recently acquired perception of God-in-Christ as determinative, in part, of his future anthropological understandings of the human person.

Sarx and soma are both employed by him over a wide spectrum, in reference to the whole person: soma has no immediate Hebrew correlate, while sarx does not necessarily equate to flesh or nephesh. While soma can refer to self, bodyness and relationality (or corporateness, as in 'the Body of Christ'), sarx denotes the weakness of the natural person, while connoting its propensity to temptation and mortality. According to Dunn, soma refers to a person *in* the world, while sarx is more suggestive of a person *belonging to* the world.[12]

The next pairing of difficult words involves psuche, and pneuma. Paul uses the former on thirteen occasions compared with 146 occasions for the latter.[13] For Paul, psuche is equivalent to nephesh (OT) and psuche in the Septuagint (LXX). Although psuche may be construed as 'immortal soul' if read in a Platonic sense, it seems more likely that Paul's concern was in expressing the reciprocal interplay between humankind's, and God's, spirit.[14] Hence, his emphatic usage of pneuma in underpinning the apostle's determined movement away from any concept of 'soul' as separable from body and the ideology pertaining to immortality. Thus *soma pneumatikon* is a Godly spirit-embued person, while *soma psuchikon* is a this-worldly engaged individual.[15]

In Paul's anthropology (following Dunn's analysis), living beings reflect embodied and relational (soma) attributes with the quality of judgement and discernment as part of mind (nous) with a natural tendency towards the spirit of God (pneuma) as opposed to non-spirituality (psuche), but endowed with a fleshly body (sarx) redolent of mankind's intrinsic frailties, open vulnerability to temptation, and ultimate demise through bodily corruption and death.[16] So Paul's approach seems to be Hebraic rather than Hellenistic, recognising 'that the terms body, spirit and soul are not different, separable faculties of man but, in fact, different ways of viewing the whole man.'[17]

Second, and away from Paul himself, the non-Pauline documents provide little material on anthropology.[18] One suggestive reference to a dualistic concept is Mt 10:28: 'Do not be afraid of those who kill the body (to soma) but cannot kill the soul (tēn psuchēn...). Rather, be afraid of the One who can destroy both soul and body (kai psuchēn kai soma) in hell'. John Cooper concludes that this pericope, in its talk of a body and a soul, cannot be referring to the same entity and therefore articulates dualistic thought-forms.[19] Wright, however, cautions against platonistic readings: that is, the 'point of telling people not to be afraid of being killed is... *if* there is an afterlife beyond bodily death', or, fear God if he does *not* resurrect you.[20,21]

Revelation 6:9 also needs consideration in reference to the verse: 'I saw beneath the altar the souls (psuchas) slain on account of the word of God...'. This, in direct parallel to Temple slaughters, suggests that the martyrs' lifeblood was poured out, indicative of bodily death as the life force drained away (Dt 12:23; Lev 17:14), rather than discarnate 'souls' awaiting resurrection.[22]

Despite such a brief review, I nevertheless conclude for the purposes of my theme that few references to a Hellenistic dualism are offered in the New Testament. Rather, humankind continues to be regarded as a unitary construction of a fleshly body and soul. The difficulties begin in regard to resurrection – are we bodies or souls, and what persists and what perishes?

This question is resolvable, as discussed at length, in the final section below.

The Degradation of Concepts of Person & Personhood

Since the eighteenth century, the concept of a person has been progressively eroded. This has been due to the emergence of more recent ideas derivative of Locke's 'Essay on Human Understanding', and which for some, are insufficiently grounded – whether anthropologically, sociologically or theologically.[23] Locke defined 'person' as merely a conscious self, exhibiting reason and reflection which were extended through time – a principle which recent commentators such as Singer, Lockwood, Tooley and Harris have further developed and elaborated into their own ethical proposals.[24,25,26,27]

In today's commerce, the concept of a 'person' has now become reduced to a locus of conscious rationality, memories, and future intentions. This thin, two-dimensional caricature excludes those in society who ought to be regarded as its rightful participants and collectively recognised – the weak, disabled, frail and elderly. Their expulsion from the fully able-bodied, so-called 'moral community' engenders the unacceptable conviction that end-of-life victims, for example, are socially and financially burdensome, and thus disposable. Additionally, immortality has now been erased from its curriculum. This trend has, it might be conjectured, throughout society gradually led to the repudiation of people whether grouped; as seen in the abuse meted out in care-homes or hospital wards towards the infirm, or as single individuals, and who so easily can become regarded as expendable prey, whether commercial, social, political, sexual, or even as 'fun' objects. We need look no further than our daily newspapers to find evidence of that disdain within society towards others, especially the young, weakened, frail or disabled – physically or mentally.

The problem with such abstract definitions of a 'person', apart from their inapplicability to the foetus, neonate, child or elderly senile, is their failure to recognise the quintessential aspect of personhood and of society: namely reciprocity and the social interactions and cohesiveness necessarily deriving therefrom.[28] Although empathic conduct is a major factor in securing the ideals of reciprocity, there are many other behavioural traits which contribute to those ends as well. But it is also useful to extend this trend in thinking to consider why autistic subjects – whatever their functional achievements are – do not behave as thugs within society despite lacking notions of empathic responsiveness, compared with psychopaths who, while often displaying considerable social charm, nevertheless dispense a considerable lack of empathy towards their victims.[29] But that merely criticises definitions. There is much more to being a person, related to createdness, and the subtle distinctions between 'what-ness' and 'who-ness'.

The Frustrations of Being Created, 'What-ness' & 'Who-ness'

With those brief introductory remarks, we come to the nub of our quest: defining what a person should be, compatible with an acceptable theologically-conditioned anthropology of mankind, as posed by the questions of Zizioulas, 'Who is man?' and 'How should he or she be?'[30]

This is fundamentally distinct from 'What is man?' which exposes its deficiency of defining mankind solely in terms of appearances, possessions and functions, that is 'what-ness': for example, 'He's the engineer with a Jaguar car', or 'She's the nice brunette in the college library'. Such epithets apply severally, neither catching the essential particularities of 'who-ness', nor even comprehending what it entails; even worse is having the name 'John Smith'. The same applies to me, since there are 192 different forms of the name 'Michael Marsh' on the internet. However, none of those include me, which can, in fact, only be distinguished by the abbreviation 'N' – a middle name based on my maternal grandmother's family of origin. In this case, I would be identified merely by the letter 'N' – which is as inanimate as is reference to my address or my telephone number!

Indeed, and in contradicting social, physical, anthropological, psychological or other approaches, a definition of a special, individual human person is not possible through use of this-worldly tokens alone. Moreover, without such background we cannot effectively get to grips with the dilemma, because of the two restrictive problems of being a human: (i) creatureliness, and (ii) the difficulties pertaining to the birth-death cycle.

(i) Creatureliness – to start, I refer to an account of the evolutionary features of mankind made several years ago by the noted biologist, Julian Huxley:

> *Evolution is seen ... as an enormous number of blind alleys. The goal of the evolutionary maze is ... a road which [leads] indefinitely*

onwards. There is but one path of unlimited progress ... the course of human evolution is as unique as its result – conceptual thought.[31]

Here we should notice, before the modern era of overt secularism, that this noted humanist saw the evolution of mankind as something special – possibly unique. Of course, the Christian believer might detect the hand of the divine in such progress and hence (as, for example, with the Victorians) that mankind's evolution was something very special – even directed. If special, then it requires deeper analysis since mankind is not unique by virtue of being a rational 'conceptualising' animal – Darwinism firmly denies that. Humankind exemplifies far more than 'mere' rationality. So, apart from the control of fire, use of language and its deployment in other abstract cognitive activities, the teaching and transmissibility of culture, what added features provide a sufficiently basic framework with, and from, which to evaluate mankind's who-ness?

Despite our createdness, we do not remain in that implied lowly circumstance. Humankind, despite creaturely contingency and finiteness, has progressively revealed aspirational or transcendental qualities (psuche) in attempting to escape from basic nature, or fleshly substance (substantia; ousia) that *ab initio*, is provided parentally. That escape is evidently exemplified through mankind's interrogations of the universe, the manipulation of technological resource, progressive understandings of genetic and biomedical process and which, in all their parts, have collectively contributed to the rich history of our species through time. While that expansive curriculum is extraordinarily impressive, there is more. Humans do acquire personhood – through relating to, and with, others – human as well as divine: that is, in relationship.

However, it is suggested that true, *ontological* Personhood (Lat: persona: Gk, hupōstasis) is only possible, if limited during one's earthly cycle, through communion with the Godhead. We thus have a 'duality' of being-ness although strictly confined within the 'psychophysical' unity of the bodily self. In this book chapter, I profess no neo-Platonic 'soul' as guarantor of immortality. True Personhood is configured in relationship and not by autonomous independence, thus implying an ecstatic (*ek stasis*) movement towards the other, in communion (ie. fellowship, or love).

In that action, mankind according to Macquarrie is 'transcended' beyond the very boundaries of self, but extended by Pannenberg as 'beyond the experience of the world'.[32,33] In the act of attempting to achieve 'freedom' (despite many contingently imposed restrictions conditioning it), a person exceeds what he was, through the process of encountering others and turning to them in love. Pannenberg continues:

> Human beings are persons by the very fact that they are not wholly and completely existent for us in their reality, but...remain[ing] concealed in the totality of their existence. A person whose whole being we could survey and...every movement we could anticipate would thereby cease to be a person [and be] treated with contempt...because human beings are in fact also existent beings.
>
> The origin of freedom lies in someone or something other than the self...basically by turning in love to others. The other person, to whom I can turn and so obtain freedom, needs to be set free in order to become himself – for he too does not derive his freedom from what he already is or was.
>
> Since in the end my fellow man is ultimately dependent as I am on the gift of freedom, the 'thou' of my fellow man cannot be the ultimate basis of freedom.[34]

Whatever quality that freedom has, space is required for inter-relational dependence: that is, to be related to – but also to be *other* than – either those upon whom, or that upon which, we depend.[35] Definitions of persons require the absolute need for relationship. But, however, in our humanly-based relationships those with whom we relate are also created finite as Pannenberg stresses, and thus ultimately unable to fully provide that true reciprocity striven for. In other words, humankind's own transcendence is unable to resolve the issue 'Who are we?' The second problem of humanity is expressed by the birth-death cycle and related issues.

(ii) The Birth-Death Cycle – createdness, of necessity, conditions humankind to the birth-death cycle. Death in its finality threatens nothingness or non-relationality, incurring the loss of each specific life history reflective of that person's individual existence and all those relationships forged during its evolution. In death there can be no more becoming since there is a cessation of being. And that would presumably be the view of the secularist who, in believing that the world comes about by chance, must conclude that nothing further obtains – nor can be expected – beyond death.

Yet contrary to that secularist view, there nevertheless co-exists the biblically perceived hope of a future conditioned by the death, resurrection, and ascension of Jesus. Through Jesus' accomplishments, humankind's temporality is extended into the future by that hope. In its temporal existence, humankind has a past and present, which considerably influences the past: 'every present is derived from the past'.[36] Moreover, the present antici-pates a future: we live for the future and in doing so, plan for it. That history – comprising everyone's past and everyone's future – resides in the present. And that evolving history represents the time allotted to each of us as we construct that history. Time is the formal ontological structure of the historicity of human existence. From that, two principles derive.

Firstly, that in death, there is no longer an end: beyond death, there is the guarantee of eternal life in the Godhead in the company of the Communion of Saints.

Secondly, personal contributory histories become a part of the totality of history, which is God's history, and, since God is eternal, all earthbound existences acquire an eternal past and an eternal future.[37] Here, rightly, I think Berger behoves us to note carefully the annals of all cultures through time, so to detect the marks of transcendental influence.[38] Viewed theologically, history ontologically becomes part of God's created world, not merely 'history' as a secular account of human endeavour throughout time. Moreover, our hope of a future life, as Webster has asserted, becomes an ontological event too, on which Jungel comments further that our past cannot be dead, but is a history articulated in and by God: 'Then, along with those aspects of our lives which have remained hidden to ourselves and to others, we ourselves shall stand revealed'.[40]

Earlier, I expressed the claim above that in mankind's own humanity, the 'Who?' question could not be resolved through earthly relationships since we are all created and thus subject to those contingent outcomes of createdness. In order to resolve this dilemma as well as the problems of freedom and relationship, in the acquisition of true 'personhood', we need a relationship with, and in, an Other.

Finding True Personhood & 'Who-ness': The Baptismal Hypostasis

In becoming a true Person (capitalised to distinguish from philosophical or other concepts of person), we need a relationship with somebody who is uncreated and thus immortal, as exemplified by the Godhead, whose constituent hypostatic Beings are defined through their true inter-relatedness ('perichoresis') as expressions of divine Love within the holy Trinity.[41] Despite our creaturely origins and the imperfections of the flesh, the answer in gaining true hypostatic Personhood is to be called through, and to be received with, divine grace.[42]

Initially, this is achieved through baptism, which is not merely a ceremonial washing away of sins, based on 'Fall' doctrine. The Fall represents a misuse of humankind's God-given freedom to have respect for the world and for others, and we don't have to look too far for continuing confirmatory evidence. Baptism, additionally and more importantly, affirms who we are by the Godhead, in continuity, both in anticipation of our dying (Lk 12:50; Mk 10:38-40; Rom 6:3-4), but also through offering to us a 'new birth from above' (Jn 3:4-7) and thus being freely lifted above ourselves to true freedom through a hypostatic (=Personhood) existence – a 'new creation' (2 Cor 5:17), and distantly analogous to the relationship between the Father and Son.[43] Through baptismal 'dying' and 'rising' with Christ we thereby become 'adopted as children' into that new relationship which entitles us to call the Father, 'Father' (Abba, Rom 8:15-17) in whom 'is perfect freedom', and to whom there is entitlement of saying 'Our Father in Heaven'.[44] It should be noted that these liturgical and biblical expressions only make sense in a metaphysical construct: they are completely unrelated to the physical world and being a (dependent) creature.

That is why baptism represents a non-repeatable, once-in-a-lifetime event. Moreover, only when viewed in that way do the relevant biblical and liturgical sayings become intelligible in conveying their rightful meanings. Ultimately, for any human, true Personhood can only be grasped through an ecstatic and 'hypostatic becoming' in Christ and in communion with the Father. It is in that relationship alone that the underlying true definition of a Person lies, as someone who now becomes, and who is therefore entirely unique, free, and unrepeatable – and within the scope of this ontological language – hardly definable by earthly symbols alone.[45]

In passing, we might take note of the fact that the ultimate un-knowability of the person recalls the apophatic traditions of the early church, in that as spiritual beings the indwelling of the godhead in us, at base, is that presence which renders each one of us incapable of being fully revealed and known. Our visible human characteristics and qualities, activities and achievements, largely characterise earthbound 'what-ness'. Only in true hypostatic Personhood do they become solely mine with a claim to absolute uniqueness, thus characterised by 'who-ness', or 'theosis' as understood by the Orthodox community.[46] This constitutes a very lofty envisioning of the real meaning of Personhood, thereby illuminating the paradoxes of living the earthly life, of being human and yet also of becoming, and being 'divinised'. Moreover, in eucharist we continually experience the communion of all the faithful on both sides of the divide – the church 'militant' (here) and the church 'triumphant' (above) within the Communion of Saints. Through Christ the Logos, in the Spirit, there is a *gathering up* of all the faithful, from all corners of the world, into divine relationship with the Godhead through the Eucharistic community (manifested by the church), and eschatologically (the Communion of Saints) in the expectation of final Resurrection of the body.[47]

Nevertheless, unfortunately, another problem arises – the paradox of presence-in-absence. We may have already noted the hiddenness in all of us, suggesting that the Delphic exhortation in Apollo's Temple – 'Gnothi Seauton' (Know Thyself) – was being somewhat over-enthusiastic with the truth. A critical restriction on self-knowledge is that I can never know other persons in their entirety, nor they me (and see Pannenberg above). People cannot be easily defined or framed, and this should worry us. It means not only that I can never ultimately know *myself* while on earth, nor *those* whom I love and are nearest to me – parents, a spouse, or my children. And neither can they me. For that reason, such knowledge as we do have so often dissolves into mere non-definitional 'what-ness'. It is this which, in part, determines the 'tragic' aspect of creaturely existence, often creating recurrent feelings of sorrow, abandonment and even total loneliness.

The exemplary 'tragic' paradox of life manifested by presence-in-absence is keenly felt and experienced deeply, passionately, and unbelievingly when the expectant and long-awaited appearance of the beloved at a train or airport terminal fails to materialise. A similar presence-in-absence arises during the viewing of a painting which although assembled by a person – is nevertheless created within creation and thus is contingently dependent on pre-existent materials. We, as humans, cannot create *ex nihilo*, which radically frustrates creaturely existence. We are entirely conditioned by what is given so that in achieving anything amounting to godlike freedom and power, we can only kill ourselves, as Zizioulas indicates in quoting Dostoevsky's Kirilov.[48,49]

This frustrating inability to catch anything forever in this life, or to get anything everlasting from its intrinsic createdness, is illustrated in the autobiography of the famous Oxford don, C.S. Lewis.[50] In earlier years he was enthralled by his so-called 'stab of joy' – especially through Wagner's music and Germanic and Nordic folk literature. But the shine wore off as he repeatedly attempted to return to these earlier sources of enjoyment for continued inspiration. He slowly learned that there had to be more, but a 'something' ultimately unattainable through the structures of this world. Finally, however, he became the world's most 'dejected and reluctant convert' in returning to the faith. At last, as he implied, he had finally been check-mated by the Grand Master.

Frances Young once wrote: 'For the most part, our experience of God is of partial presence or apparent absence...' while Jesus' acute sense of God's absence was all too disturbingly real (Mt 27:46).[51]

God, and humans (you and I), are unknowable. And yet it is through belief, with the renewed baptismal hupostasis of being reborn from above, together with the ontological hope of life everlasting, that ultimate rest (stasis) is obtainable within that paradoxically unknowable Godhead. Then the blindfold will be lifted and we shall then know who we are, and for evermore.

* * *

1. Calcagno JM, Fuentes A, 2012. What makes us human? Answers from evolutionary anthropology. *Evol Anthropol* 21: 183-194.

2. Gibson J, 1998. *Language and Imagery in the Old Testament*. London: SPCK 37.

3. Martin-Achard R, 1960. *From Death to Life. A Study of the Development of the Doctrine of the Resurrection in the Old Testament*. London: Oliver & Boyd [ET: J Penney-Smith], 3ff.

4. Robinson J, 1957. *The Body: A Study in Pauline Theology*. London: SCM Press 1957, 11.

5. Johnson A, 1964. *The Vitality of the Individual in the Thought of Ancient Israel*. Cardiff: Cardiff University Press 37ff.

6. Johnson A, 1961.*The One and the Many in the Israelite Conception of God*. Cardiff: Cardiff University Press 1-2.

7. Johnson 1961, 8; Wheeler-Robinson H, 1981. *Corporate Personality in Ancient Israel*. Edinburgh: Clark. And also Num 21:4 'And [Israel] journeyed from Mount Hor ... and nephesh-ha'am ... became impatient on the way'. The collective noun for the people (body) of Israel here is singular.

8. Stacey D, 1956. *The Pauline View of Man, in Relation to its Judaic and Hellenistic Background*. London: Macmillan 26-27.

9. Stacey 1956, 222.

10. Dunn JDG, 1998. *The Theology of Paul the Apostle*. Edinburgh: Clark 54.

11. Rom 11:1; Phil 3:5-6; Gal 1:14.

12. Dunn, Paul 72-78.

13. Stacey, Pauline view 145.

14. Stacey, Pauline view 121; 1Cor 6:17 – 'But he that is united with the Lord is one with him in spirit' (èn pneûma).

15. Dunn, Paul 78; Dahl, ME, 1962. *The Resurrection of the Body*. London: SCM 15.

16. Stacey, Pauline view 199.

17. Ladd, GE, 1975. *A Theology of the New Testament*. Michigan: Eerdmans 457.

18. Wright NT, 2003. *The Resurrection of the Son of God*. London: SPCK 401.

19. Cooper 2000 117.

20. Wright, Resurrection 431.

21. Cullman O, 1967. Immortality of the Soul or Resurrection of the Body. In: Stendahl Krister (ed), *Immortality & Resurrection*. New York: Macmillan 27.

22. De Silva L, 1979. *The Problem of the Self in Buddhism and Christianity*. London: Macmillan 83; 168-note 36.

23. Locke J, 1869. *An Essay Concerning Human Understanding*. New York: Prometheus Books 246-250.

24. Singer P, 1979. *Practical Ethics*. Cambridge: Cambridge University Press 76.

25. Lockwood M, 1985. *Moral Dilemmas in Modern Medicine*. Oxford: Oxford University Press 10.

26. Tooley M, 1983. *Abortion and Infanticide*. Oxford: Oxford University Press 123.

27. Harris J, 1983. In vitro fertilisation: The ethical issues. *The Philosophical Quarterly* 33: 225.

28. Gordijn B, 1999. The troublesome concept of the person. *Theoretical Medicine and Bioethics* 20: 347-359.

29. Marsh MN, 2013. Empathy: Mirroring another's predicament – or more? *Antonianum* 3: 409-430.

30. Zizoulas J, 1991. On Being a Person: Towards an Ontology of Personhood. In: Schwöbel C & Gunton CE (eds), *Persons Divine and Human*. Edinburgh: Clark 33-46.

31. Huxley J, 1941. *The Uniqueness of Man*. London: Scientific Book Club 9-15.

32. MacQuarrie J, 1983. *In Search of Humanity: A Theological & Philosophical Approach*. New York: Crossroad 25-37.

33. Pannenberg W, 1972 (1962). *What is Man?* Philadelphia: Fortress Press 10-11.

34. Pannenberg W, 1971. *Basic Questions in Theology* (Vol 3). London: SCM Press 112-3.
35. Gunton CE, 1991. Trinity, Ontology and Anthropology: Towards a Renewal of the Doctrine of the *Imago Dei*. In: Schwöbel C & Gunton CE (eds), *Persons Divine and Human*. Edinburgh: Clark 53.
36. Jungel E, 1975. The Death of Death. In: *Death – The Riddle and the Mystery*. Edinburgh: St Andrew's Press 117ff.
37. Idem, Death 119-120.
38. Berger P, 1969. *A Rumour Of Angels*. London: Penguin 70;81;86-88;99-100;104.
39. Webster J, 2005. Hope. In: *Confessing God: Essays in Christian Dogmatics II*. London: Clark International 195-214. Also Berger, Rumour 80.
40. Jungel, Death, 121. Also Miranda J, 1977[ET]: *Being & The Messiah*. Maryknoll: Orbis: 'all our actions ascend to the knowledge of God in Heaven and there their reward accumulates like a treasure', Lk 6:23; Matt 5:12; Acts 10:4 etc: 58.
41. Clément O, 2000. *On Human Being – A Spiritual Anthology* [ET]. New York: New City 28.
42. Second Collect for Christmas Day: '....grant that we, being re[-]generate[d] and made *thy children* – by *adoption* and *grace* – may daily be renewed by the Holy Spirit ....' (my insertions and emphases).
43. '... the believer becomes a new being, a being in God'. In: FM Braun, 1960. *Jean Le Théologien* Paris: Gabalda 129. Also 2 Cor 5:17. Again, the 'new birth' [given] in baptism by water and the Spirit (Common Worship Confirmation Service).
44. Pannenberg W, 1996. Baptism as remembered 'Ecstatic' Identity. In: D Brown & A Loades (eds), *Christ: The Sacramental Word*. London: SPCK 77-88.
45. Macquarrie, Search 26-7.
46. Zizioulas, On Being 43-44.

47. The Greek word 'Logos' and its verbal cognate 'lego' can imply a gathering up, as mentioned by Heidegger (*Introduction to Metaphysics*, 1959, 170ff). Note the Blessing for Christmas Day: 'May He [that is, The Logos], who by his Incarnation *gathered* into one, things earthly and heavenly...'

48. Arendt H, 1959. *The Human Condition*. New York: Double Day.

49. Zizioulas, Capacity 432.

50. Lewis CS, 1977. *Surprised by Joy*. London: Fount Books 20, 136 ,176, 182.

51. Young F, 1985. *'Face to Face'*. London: Epworth Press 62.

Chapter 7

Ascribing Moral Status
to the Human Embryo-Foetus

The previous chapter, in its approach to the difficulties surrounding human dignity, derived its material from theological perspectives seen from within a created universe. I now proceed to consider, in more depth, the possible reasons why the embryo-foetus should not have a dignity apportioned to it and thus to be considered of moral significance.

In approaching this subject, I am not necessarily concerned with frozen embryos, or embryos created *in vitro* purely for scientific study. Neither am I asking whether the embryo-foetus is deserving of a respect 'lower' than that accorded to adults.

Essentially, this chapter offers the claim that biomedical understandings contribute important additional insights to this subject while not obviously exhausting other relevant perspectives. My argument is that unless we can afford moral respect to the embryo-foetus – and that is a topic which has lost its incisiveness largely, I claim, through abortion practices – neither can we attribute moral regard to babies, neonates, children or those whose flourishing is impeded by disability or some other kind of dysfunction. In other words, we either apportion dignity and moral regard to all members of the human race – including the embryo-foetus, or we give up trying to have any ethical principles.

To what extent, then, could the human embryo-foetus be said to be 'morally considerable' and by what criteria should the embryo-foetus be evaluated? Goodpaster's phrase 'morally considerable' stems from GJ Warnock's book in which he outlined criteria for moral status as experienced from the other side, that is, from the recipients of ethical practice.[1,2] From that perspective, what kind of claim(s) can the embryo-foetus make for itself, so to speak, thus to be reckoned morally considerable by other rational (and hopefully moral) agents?

The Legal Background

It is likely that in the public mind the provisions of the Warnock Report (1984), later enshrined in the Human Fertilisation and Embryology Act (1990), will be in view.[3] That statuary pronouncement reduced the latest permissible time for effecting 'therapeutic' abortions from 28 to 24 weeks gestation. Second, it limited experimental research on living human embryos to the 14-day stage only, based on two criteria: the occurrence of the embryonic primitive streak, and twinning.

Nevertheless, save for the conditions governing embryo research, the embryo-foetus has, surprisingly, no protection in UK law.[4] This point has been recently emphasised in the British law courts (The Court of Appeal). The victim is a 7-year-old child, born of a mother who when pregnant drank so excessively that this girl now has foetal alcohol spectrum disorder. The appeal against 'grievous bodily harm' inflicted on another person (mother against child) was discharged on the very grounds that a foetus is not regarded as a person, and therefore has no legal rights of redress, at law.[5] The *moral* claims of the embryo-foetus, if any, therefore depend on (i) what claims can be made on its behalf within society and (ii) what value or significance society places on each developmental stage of pregnancy.

Vocabulary & Some Definitions

Apart from legal issues, there remains the problem of vocabulary and its applicability to the embryo-foetus. The reproductive process, as discussed more fully below, results in the specification of a new human individual. Here 'individual', used as a noun either means, or refers to, a single human being.[6] But much discussion concerning the early stages of the embryo-foetus, from zygote (arising from fusion of sperm with egg) onwards and highly relevant to its perceived moral, social and legal status, involves the words 'person' and 'personhood' and the questions 'Was I ever a zygote?', or, 'Can a zygote be a person?'[7,8,9]

But can the newly-formed zygote, as sperm and ovum unite, sensibly be regarded as a 'person' or something with 'potentiality' or 'personhood'? And, dependent on the answers given, at what stage in its development could moral considerability be applied?

First, my major concern here is that these questions fail to define the issue, and that they are inappropriate formulations applicable to a zygote or later intra-uterine object. They simply do not encourage effective, rational thinking on the subject. The one-cell zygote acquires the full and unique DNA complement (or genome) of a distinct human individual, which, in my view, merits full consideration as a member of the species. This is the means through which every individual is created, developed and delivered into the world – there to manifest the onward continuity of growing, maturing, and functioning.

Second, we encounter the difficulty aired in the last chapter – that there are no uniform definitions of 'a person'.[10] The word has succumbed to becoming a descriptive catalogue of properties presumed to frame competent adult humans only – a procedure that is 'logically confused and morally objectionable'.[11] Much philosophical reflection has rendered the idea of a person similar to a totem pole to which can be attached items seen by each writer as demonstrative of a mature, moral personhood, such tick-box approaches based on stipulative attributions failing entirely in positing the individual within either a sound anthropological, sociological, or theological footing.[12] At what age are these arbitrarily-defined properties acquired? Do they not change with the continually altered contingencies of life, of living? And more fatally, of growing old? Nor are these attributes in play throughout any individual's life.

Importantly, these shopping lists woefully exclude the embryo-foetus, as indicated by GJ Warnock. That a deformed foetus is seen merely as a source of tissue or organ transplant material, or that an adverse 'genetic diagnosis' calls for immediate abortion, hints at eugenics under another name and the re-emergence of terms like unfit, incapable or even undesirable.[13] Pliable definitions of persons, chosen to be wide or restrictive, are employable instrumentally across committee tables or public places for launching, or even insisting upon, agendas comfortable to the proposer but deployed as political tools inimical to others' viewpoints.

Third, another obstacle to clarity involves the notion of 'potentiality', as in 'a potential person'. Can this mean a potential in being able to develop to maturity – to be; to do; or merely to aspire? And the answer, boringly, seems to be: 'so what?' Surely, living one's life inevitably involves potentiality, the balance between probabilities and possibilities, between our innate and acquired capabilities and capacities, the opportunity for their realisation and development, and what will not, or might never be achieved.

Much of this implies and requires a competent agent, so there is little point in conferring the idea of potential to an embryo-foetus, a child with birth-induced brain injury or manifesting the long-term consequences of a metabolic degenerative condition. Indeed, the potential of all (normal) embryo-foetuses to become competent adults is a truism, merely dependent on their early developmental stages and the outcome of biological, genetic, molecular and environmental contingencies throughout their earlier phases of existence.[14]

Biological Considerations
We should be reminded that (i) the specification of new individuals through fertilisation by sperm of a mature human ovum endows them with a *unique* genetic foundation, and (ii) the

process of speciation thereby perpetuates in continuity the reproductive property of living matter. Both cellular contributors to the process, ovum and spermatozoon, are themselves living cells, albeit specialised in function and yet functionless without amalgamation, of limited lifespan, and containing only one-half the normal human genome (that is, 24 chromosomes each – 23 'autosomes' together with either one X, or one Y, sex chromosome). They arise from individuals who, in their turn, were derivative of the same creational process. The begetting of any new human individual with a full complement of 48 chromosomes, once sperm and ovum have united, illustrates the idea of continuity. By that I mean being incorporated into, and thus becoming part of a sequential, unbroken seam demonstrative of the reproductive capacity of living matter evolved from its inorganic base. Life is not created on each moment of procreation: the product is a new, living, and unique individual.

It might be claimed that the prevailing view within UK society is that abortion is permissible and freely available until the 24th week of pregnancy. Although inconsistent with the Abortion Act, that perception provides a useful watershed since, after that time, the embryo-foetus is usually accorded some degree of moral status. In the next section, I evaluate that idea more fully.

On Being *In Utero*: The Changing Landscape

There are various ways of evaluating the conceptus *in utero*: its emergence from the shadowy realm of the womb; the development of foetal competence; and the influence of maternal lifestyle.

Firstly, the Foetus In Utero: it is evident that 3-D foetal ultrasound (US) has now revolutionised our ideas, bringing to the public mind a large repertoire of observed activities. These activities render the foetus a subject of some wonder, if not awe, whatever their rightful functional interpretations. Routine antenatal US encourages a deeper relationship between parents and the developing baby, offering possibilities of naming. Rather than disregarded as a passive parasite within the mother's womb, the embryo-foetus is now seen to be a complex organism developing in continuity with its intrauterine environment, and, as pregnancy progresses to term, in continuity with its future environment beyond the womb.[15]

Yet old habits die hard. Clare Williams refers to differing vocabularies exemplary of positions established c1967: 'pro-lifers' speaking of baby or unborn child, while 'pro-choice' campaigners employed foetus, 'it', or pregnancy, thereby distracting attention from any hint of an individual human, of something appealing – even loveable.[16]

But four aspects arise here.

First, such antithetical stances (pro/anti-abortion) have been made to yield, I would suggest, to the thrusting status of the developing conceptus as exemplified by the enormous expansion of premature baby units in major hospitals. These units have emphatically drawn attention to the fact that a foetus, scarcely 20+ weeks gestation can be cared for in an incubator and ultimately given opportunity to be taken home in mother's arms. Note the fine line here between the medical opportunity for foetal survival and the lawful opportunity for foetal destruction (usually within 24 weeks and not beyond under normal circumstances). The viability of the foetus burgeons clearly into view here. If there is concern for foetal viability, and that surely must be the underlying concern of all those who staff and offer their expertise in these units throughout the country, then surely there must exist a moral concern for the life and survival of each of these early foetuses.

Second, moral support is offered by the Stillbirth (Definition) Act (1992) defined as a spontaneous intra-uterine death (but only) at 24 weeks gestation or beyond which must, by law, be registered. This allows subsequent burial or cremation and opportunity for respect and honour to be given to that individual as a rightful member of the human race and as the cherished offspring of its parents. Such legal enactment further enhances the moral claims of the foetus, and indirectly helps to re-emphasise continuity – drawing attention, and giving point, to such an early intrauterine demise as a phase recognisable in the longer, time-related processes of achieving infancy, childhood, adolescence and subsequent evolution to mature adulthood.

Third, remedial procedures, clinically applicable to the foetus are being widely developed. This expanding intra-uterine thera-peutic landscape has vouchsafed to the foetus the status of 'patient'. Yet this ascendancy has generated other problems relevant to such a prioritised status. Does the foetus, as *patient* automatically qualify for full recognition as a human being, and, indeed, a *morally* reckonable human being? Some disagree, recog-nising – albeit grudgingly – that while a procedure may be required, the recipient is not a proper human being.

Nevertheless, I am unable to follow such argument. Treating a foetus is surely no different from treating brain-damaged infants or even older malfunctioning children or adults who, although incapable of consenting, become 'patients' in need of relief from a self-evident clinical problem. It also follows that an even greater sense of value would ensue if the gender of the foetus were already known, since it might well be referred to as 'he', or 'she', or even to a specifically named individual, rather than merely an 'it thing'. In such circumstances, we already have an 'I-Thou' interaction because of the caring relationship now in place with that patient, and, if a human patient, then a morally-consid-erable human with a right to treatment. Moreover, that right to treatment brings entitlement to full medico-legal responsibility and duty of care accorded other human individuals by the medical and clinical practitioners involved.

Fourth, the attribution of respect to the foetus pertains directly to its capacity to experience pain from manipulative interventions whether intra-uterine or within the intensive-care incubator. Foetal responses to painful stimuli (as seen, I stress, from an adult perspective), such as a needle prick, result in motor withdrawal behaviour and incremental rises in 'stress' hormones. Thalamo-cortical connectivity, initiated ~14-15 weeks but incomplete until ~25 weeks gestation, is essential for transmission of sensory messages, of whatever type and originating peripheral locus, to the cerebral cortex.[17] Data are lacking on the gestational time at which the foetus is competent to consciously perceive varied incoming sensory ('afferent' qualities). That is, how the relevant experiential learning could have taken place, if it does: and, of course, in the absence of a verbal account from the foetus, we cannot be sure. Over time, a broad consensus has emerged that the foetus does experience pain, so that appropriate anaesthesia is necessary: if animals during surgery require protection, so do foetuses.

Secondly, Foetal Competence In Utero: here, I briefly review evidence for some of the competences demonstrable during foetal growth. Newer technologies are beginning to uncover and elucidate the many perceptual, behavioural and learning tasks acquired by the foetus as it interacts with its environment, and the factors which disrupt them.[18] Here we have the foundations of a new 'foetal neurobiology'. There is one functional aspect of importance: the development of auditory function in the mid- to late-term foetus. Newborns show preference for their mother's voice, react to the prosodic rhythmicity and contours of their native language, and respond specifically to music played to them *in utero.* These observations again emphasise the *continuity* of the foetus through birth and into the immediate post-natal period with capacities that bind it specifically to its mother.[19] This facilitates the relationship, through voice, to the mother's face, which the foetus would not have seen previously. Generally, the foetus has an innate ability to discriminate between real and fake faces, preferring its mother's face to that of a stranger even when other variables are confounded such as hair colour, body odour and skin pigmentation.

Thirdly, Epigenetic Factors Influencing Development – Even Beyond the Womb: we need to examine critical data that are relevant to the growing foetus in utero, but also to other environmental (epigenetic) factors which play critical roles in what we become. It is easy to list the prohibitions – as well as social opprobrium meted out to women during pregnancy.[20] Such prohibitions include smoking, alcohol (leading to the foetal alcohol withdrawal syndrome and its longer term sequelae), recreational drug taking (and 'cold turkey' in newborns acutely withdrawn from morphine exposure), and pharmaceuticals (thalidomide). Other environmental dangers include radiation (Chernobyl) and its long-lasting gene-mutational influences, very severe malnutrition and its effects on the developing embryo-foetus (the civil wars in Africa), infections (French cheese and toxoplasmosis), and the MMR (the Measles/Mumps/Rubella immunisation schedules for post-natal babies in operation throughout the UK) scare and its possible disastrous outcomes on vaccination rates for these toxic infections on the unprotected foetus. Even the effects of terrorism influence fecundity.[21]

This long list is in part reflective of the immorality of governmental and managerial incompetence; of tribal hatreds; of dispossession, rape and starvation; and imposed maternal stress, all having their inevitable toll on the foetus *in utero*, and its later physical and psychological competence during childhood and adult life.[22] The moral significance of these adverse influences is all too clear and a matter of deep regret. They also reflect the longitudinal effects of many of these catastrophic influences, again highlighting the important notion of continuity in the growth and anticipated gains, which every human being, from the moment of fertilisation to the point of death, is entitled to expect.

But we need to be very conversant with the longer-term influence of environmental factors on the outcomes for the foetus. There are three emergent areas of study into critically adverse environmental factors influencing the longitudinal expectations of the growing foetus which are of enormous importance: (i) The after-effects of the Dutch Winter of Starvation (November 1944-April 1945), (ii) The impressive, long-term prospective study of children and their parents' life-styles in Bristol, UK (Avon Longitudinal Study of Parents & Children: 'ALSPAC'), and (iii) the influence of diet in a remote agrarian village in northern Sweden.[23,24]

(i) Observations on Dutch Famine victims have revealed *epigenetic* (non-Mendelian) environmental influences persisting until the sixth decade in those individuals conceived at varying periods during that period of very harsh low (maternal) calorie intake – often less than 1,000 calories per day.

First trimester (starvation between 1-13 weeks) foetal exposure only, resulted in increased rates (x3) of coronary heart disease, obesity and hyperlipidaemia in relevant offspring. Second trimester (14-26 weeks) exposure led to obstructive airways disease (x1.6) and renal (x1.5) problems, while third trimester (27-39 weeks) exposure resulted in low birth weights (x-0.9) and diabetes (x1.4). These studies are still incomplete and will crucially be applied to the second generation children (now adults) to determine whether those epigenetic effects have been 'inherited', and incorporated into the gonadal tissues, thus to be capable of true *genetic* (Darwinian) transmission to the grand-children of those (grandparents) originally exposed. These studies, soon to be started, will bear significantly on the role of environmental factors in changing the genetic landscape *in utero* of every individual – across several generations.

(ii) Some research (ALSPAC) pertains to the slow growth period (SGP: 9-12 years-of-age) in boys. For the first time in the field of genetic epidemiology the influence, as well as the importance, of the male line has been recognised. Thus, boys commencing smoking in SGP at 9-years-of-age later fathered sons who, during *their* subsequent SGP, exhibited an increased BMI (body mass index): they were obese.

(iii) The combined Swedish/UK-ALSPAC analyses identified influential environmental effects operative through each grandparent towards subsequent offspring. These effects impinge on the intrauterine growth of gonadal tissues and cells, and other organs (pancreas, heart, lungs) at their crucial developmental trajectory and growth during gestation and beyond. More crucially, the combined studies revealed that *paternal* grandfathers experiencing high food intakes as boys (during *their* SGP) led to higher mortality rates (relative risk) in their grandsons. Analogous effects during the *paternal* grandmother's childhood caused similar increases in granddaughters' mortality. Alternating periods of low food intake reversed these trends.

Such amazing revelations point to gender-specific, transgenerational mechanisms operative on human offspring. These traits are transmitted from paternal grandparents through the male line – thence through the same *father* to either (grand)son or (grand)daughter. Clearly, these longitudinal studies have refocussed interest on (a) critical growth periods during foetal and post-foetal life, (b) the successive developmental phases (*in utero* and beyond) in the growth of ovaries and testes (as organs) and (c) the maturational profiles of ova and sperm which originate within them.

Additionally, these studies are emphatic of the varied environmental influences likely to disturb the prolonged physiological natural history of other organs – such as the developing pancreas and its role in determining any future propensity to develop diabetes: and likewise, in terms of diseases associated with the heart, lungs and kidneys. Again, the idea of continuity assumes considerable dominance in these long-lasting effects.

More broadly, these findings bear importantly on Public Health programmes relevant to the developmental well-being of children during their SGP, and thus to our responsibilities to future generations. Clearly, the SGP is emerging as a critical period of tissue and organ growth which, in deflecting sole attention away from the mother, engages grandparents and prospective fathers in *their* critical roles in the reproductive cycle, and its forward-feed generational consequences on subsequent foetuses, including their well-being *in utero* and beyond.

The moral implications of knowing, understanding and applying these new data are clearly evident, and must obviously be taken into consideration when new public health initiatives are envisioned, whether locally, nationally, or internationally.

Considering the Early Embryo-Foetus

The previous discussion has aired the many relevant factors attributable to the (undoubted) moral status of the foetus aged, say, 20 plus weeks gestation, and pertaining to its structural growth, functional capacities and neurophysiological attainments – attributes related to the continuity of its intra-uterine period with that of the post-natal period and beyond, into adulthood. We have also considered the foetus as a patient, and the subject of obligatory rights during its treatment by professional health care personnel.

But next, I turn to a far more controversial issue – the moral status of the embryo-foetus in its earliest phases, by examining the positions assumed in the Warnock Report and by Normal Ford.[25] Their arguments are based on two developmental phenomena: the emergence of the 'primitive streak', and the process of twinning.

First, some elementary biology. The fusion of egg and spermhead results in a single-celled *zygote* which immediately divides to produce two new embryonic cells or *blastomeres*. Further divisions expand the mass of blastomeres (4-8 cell stages) into a ball or 'morula' (16+ cell stages). Cavitation of the morula produces the *blastocyst* (64+ cells), which embeds in the uterine wall. The terminology is arbitrary. What is salient to grasp is that the process of individuation – from the moments of fertilisation onwards – exemplifies a continuous and defined series of genetic activations resulting in specific cellular differentiations and migrations which progressively, and incrementally, define the newly-developing individual.

The blastocyst reveals the emergence of cell groups – the 'inner cell mass' – and its segregation from other cell groups, the latter giving rise to the encompassing foetal membranes (chorion and amniotic sac). At approximately 14 days, the blastocyst prepares to embed in the uterine wall. The activities of the chorionic vessels, by their invasion of the uterine wall, initiate the developmental origins of the placenta. Following this, cells migrate along the prospective back ('dorsal aspect') of the embryo, thus defining the so-called 'primitive streak' from which ultimately arises the individual's vertebral column, and within it, the neural spinal (neural) cord.

Although the streak is one of the first major external visible features of the differentiating embryo, it emerges from a complex series of preceding interactions between a large number of genes and their products of activation. It therefore has no specific significance with regard to any moral property of the embryo-foetus. That is most important to observe. Ford thinks the streak defines the new individual (his book, p171), as does Warnock, but without providing any argument or supporting data to underpin her committee's views. Neither does she attempt to substantiate her 'moral' corollary that experimentation should therefore not proceed beyond this arbitrary biological time-point (14 days) beyond which twinning does not occur (her report, p66). But why should this be so? There is no safe connection between the streak and personhood, despite the writings of these two authors.

Next, these authors appeal to twinning, asserting quite erroneously as a result of this developmental anomaly, that there can be no 'individual' existing before 14 days post-fertilisation. But first, a few remarks about twins and twinning.

There are four types of twins.[26] 30% occur before blastocyst formation (0-4 days post-fertilisation) each having its own distinct membranes, the outer chorion and inner amnion. The majority, however, (65%) are monochorionic (MC), diamniotic (DA), arising at +4 days after the chorion membrane has formed: each embryo has its own amnion with placenta shared. In later gestation (c10 days), embryonic splitting leads to the rarer (4%) monoamniotic twins, each of the pair inhabiting the same membranes (chorion and inner amnion). At an even later stage (~14+ days) attempted division of the embryo yields conjoined twins (1%), 80% being female. Since all later divisions of the embryo-foetus are increasingly biased towards the production of female twins, the X chromosome clearly contributes to these abnormal processes.

My quest is to show the erroneous appeal to twinning as a guide to the origin of a human person. In the following, I employ aberrant (or pathological) forms of twinning to underpin my argument.

(i) Conjoined Twins: these arise very late, and attempts at splitting (~14+ days gestation) interfere with the left-right planes of embryo-foetal orientation which have already now been decided. This attempt to divide embarrasses the sharing of single, midline structures (brain, heart, windpipe and oesophagus, intestine, and genitalia). There was obviously only one individual from the outset, immediately after fertilisation was complete. There never was any ill-defined, 'nothing' which disappeared once doubling started: but that is the argument used by Warnock and Ford. In passing, we should observe another important fact which neither Ford or Warnock seem to have grasped. At this very late stage of division, there is only *one* set of embryonic membranes (one chorion and one amnion) – predictive of one *single* foetus. This is a most fatal blow to Ford and Warnock who declare that there was no individual, antecedent to the occurrence of twinning, even though this 'non-

existent individual' actually fulfils their criteria of personhood – blastocyst embedment (= placental formation) and creation of the primitive streak – and note, *before* its unsuccessfully attempted separation.

(ii) Spontaneous Multiple Pregnancies: in subjecting the position of Ford and Warnock to further scrutiny, let us examine their notion that twins ('B' and 'C', say) are *definitive* individuals. But the real possibility of a further division of one or two of these 'definitives' (B or C) is highly possible, thus to produce triplets (D) or quadruplets (E), and so on. And if this process continued indefinitely, we would be in the rather silly position of conferring definitive personhood on each succeeding pair, only for that status to be lost with each ensuing division. Such a view does not make sense.

(iii) Skewed X-Chromosome Inactivation: after conception, all cells (blastomeres) of normal female embryo-foetuses contain two X chromosomes, one maternally-derived and the other paternally-derived. Random inactivation of one of these two inherited X chromosomes must be completed in *every cell* at an early stage of gestation (by ~ 4 days post-fertilisation) otherwise development is abnormal (termed 'Lyonisation', after the recently deceased Mary Lyon who recognised this event c1960). This results in 50:50 distributions of maternal/paternal X chromosomes throughout the female embryo-foetus. If we imagine that maternal cells are pink, and paternal cells blue, every adult woman manifests an equal mosaic of pink/blue cells throughout her body. But skewing of X chromosome inactivation during the twinning process could result in one twin being all blue (containing majority of paternally-derived X chromosomes), the other all pink (containing most maternally-derived X chromosomes).

But we would not be able to discern the occurrence of that skewing anomaly without a biological indicator. Now, it has been amply shown that one of a pair of female monozygotic (MZ) twins may manifest certain X-linked (= sex-linked) disorders, such as haemophilia, or red-green colour blindness.[27] But X-linked disorders, usually, are transmitted only through clinically *normal females*, and manifested clinically only in *affected males*. That one of a female twin pair actually manifests one of these diseases *clinically* would signal the result of aberrant skewed inheritance of maternal/paternal X chromosomes.

These cases prove that there existed one female embryo before twinning occurred whose maternally-derived X chromosomes carried the disease-bearing abnormality. Because of skewed X inactivation, one twin now predominantly carries maternally derived cells (all bearing disease-trait X chromosomes: she would be pink all over), the other carrying predominantly paternally-derived cells (all bearing non-disease X chromosomes and blue all over). What is evident, from our perspective, is that there was only one original individual, with a post-Lyonisation twinning process (as early as three-four days post-fertilisation) producing one disease-bearing twin and a second disease-free twin.

We need to be clear of the implicit assertion here. There is only one female XX individual produced at fertilisation, *who continues her existence through* the twinning process. This female, originating with a familial transmission of an X-linked disease, remains the carrier of this disease from fertilisation and forwards into adult life. The disease never arose by 'chance' through a new, disease-bearing mutation occurring after fertilisation, or after twinning had taken place. But that underlies the assertion being made by Ford, who states that those cells which existed from the moment of fertilisation, up to the twinning event, have no developmental relationship to the post-twinned individuals, an assertion shown by these special examples to be complete nonsense.

(iv) Chromosomal Mosaicism Leading to Phenotypic Discordance in MZ Twins: while discordance for (sex-linked)

diseases arises from skewed X inactivation (explained in (iii) above), an analogous outcome may be occasioned by an abnormal cell division. When a cell divides, its chromosomes are duplicated and assigned to each of the two resulting daughter cells, but during this, a chromosome may be completely lost, or a segment transferred ('translocation') to another chromosome (one of the commonest translocational occurrences is trisomy of chromosome 21, resulting in Down's syndrome). The process leading to aberrant chromosomes is very common during early foetal development and is a major reason for spontaneous abortion since the disruption in chromosome numbers produces a non-viable embryo.

If a female zygote (46+XX) during or after the twinning process sustains loss of one X chromosome, the original continues as a normal female (46+XX), while the other abnormal female twin (46+XO: the 'XO' indicating loss of one complete X chromosome) discordantly manifests Ullrich-Turner syndrome.[28,29] Likewise, if a normal male embryo, 46+XY, destined to become a twin incurs loss of a Y chromosome at some stage during this process, then that abnormal body of cells also now becomes, genetically and phenotypically, a female with Ullrich-Turner syndrome (46+XO).[30] From these detailed cellular and chromosomal observations, it is not too difficult to conclude that from an original single, and genetically-unique zygote, a second anomalous individual results. The continuity of the original individual from fertilisation, and from which the second (abnormal) individual splits, is nevertheless self-evident. This is crucial information because it clearly indicates that the zygote arising at fertilisation and its progeny, up to the point of separation, *neither* ceases to exist, *nor* can be dismissed as unrelated or even as Ford would put it – 'ontologically irrelevant', simply because of the twinning event.

With those telling arguments, we avoid the embarrassing outcome of Ford and Warnock who assert that before the 'definitive' twins appeared, there was no relational continuity with the earlier collection of blastomeres that developed from fertilisation up to the point when the anomalous biological event took place. On all grounds cited, those suppositions are no longer credible and must therefore be rejected absolutely.

Where does this get us? My claim is that the use of twinning and the appearance of the primitive streak to define the start of new persons, and that their occurrences negate all the preceding stages in development from fertilisation, completely lack biological and logical foundations.

The idea of continuity applies here. That is, the continuity of the processes which, from fertilisation, properly mark the biological origins of the new human individual. To mark the streak as the commencement of special moral significance (at some arbitrary time-point ~14 days post-conception) is to miss this principle of continuity, and its vital relevance to the earlier developing embryo-foetus. Moreover, as indicated in the previous section, the principle of continuity, in its embrace, reaches back to the developing ovaries of the mother while she, herself, was *in utero* (that is, in the grandmother's womb), and to the later, crucial processes necessary for maternal ova to be capable of fertilisation. Furthermore, continuity stretches back to an even far earlier period, involving environmental influences exerted on the grandparents, two generations before, as also noted above.

The Moral Status of the Human Embryo-Foetus

I began with reference to GJ Warnock and his concerns for the objects (ends, and recipients) of moral considerability – recipients being on the receiving end of the behavioural attitudes (and moral edicts) of others presumed to be rational and moral. Geoffrey Warnock argued that morality, in general, can never be a system of prescribed rules incumbent on all to obey. Rather, through one's own experience(s) he suggests, one might *want* to fulfil certain rules for the betterment of the other. But then, who is the beneficiary of such moral action – only supposedly sensible adults themselves? Clearly, 'No' . For Warnock, the term 'person' is far too restrictive within this realm of thinking, because of its inapplicability to babies, children and those incompetent through various biological changes throughout life. But the moral claim to 'considerability' which the latter deserve, despite absence of rationality, or capacity to act and judge for themselves, is simply that they, like everyone, suffer Warnock's human predicament. We, as human beings, are 'all in it together'. And in recognising that principle, any sense of future 'potential' is thus inappropriate at that early stage, and to be thoroughly rejected. The working criterion is 'sufferability'.

Warnock adopts the Golden Rule, even if he does not specifically apply his ideas of 'considerability', or 'sufferability', as far back as the impregnated ovum. And from the previously detailed section, it has been clearly demonstrated that the new human individual originates at the time of fertilisation, and not at some arbitrarily man-made time-point later, in order to supposedly justify predetermined philosophical or political ideals.

At the commencement of this chapter, I stated why the use of person and personhood are inappropriate words for descriptive use when considering the embryo-foetus. All we can say of the early embryo-foetus is that it is a newly created human being. Neither should we deliberate solely upon the acquired characteristics of 'what-ness' as explored in the previous chapter, but rather 'who-ness' or the *one who is*. And in response to the retort '...is *what*?' comes the only reply 'The one (he or she) who is'. This open definition, *he* or *she who is*, reflects and is emphatic of, the physical as well as the metaphysical status at any, and every, moment of that person's lifelong career – from zygote onwards. Having already seen the futility of trying to force some undefinable notion of 'personhood' or 'potentiality' onto a post-mitotic zygote, or later embryo-foetus, we are cleared to evaluate the latter on its own terms, yet well within the scope of that definition.

That form of evaluation, in part, depends on the continuity of life, which originated from inorganic matter millions of years ago. Moreover, that living continuity is maintained and expressed in the gametes, which, on coming together, results in specifying the new, genetically unique human being. I have also shown ways in which that continuity is embodied in the developing gonadal organs when the parents themselves were either *in utero*, or in their later childhood years, and possibly being subject to those specific environmental influences which impinged on those parents when they were children, as well as their paternal grandparents before them.

These are the moral criteria relevant to all human individuals at whatever stage in their biological and personal evolution. We are all inextricably intertwined in the everlasting and persisting nexus of relationship and continuity. Continuity and relationship embrace the evolutionary cohesion between the new individual and its parents and grandparents, thus to mould the phenotypic expression of that individual as it grows into adult life. And, as part of that relational matrix in which families over at least three generations are so intimately intertwined, the new individual deserves moral respect: it cannot be divorced or dissociated from those continuous genetic, environmental or societal influences. It is that background, I believe, upon which and from which the necessity for abortion, and experimentation on embryonic life, should be based.

In passing, it should be noted that scant reference to these important considerations rarely surfaces in the numerous philosophical discussions of 'personhood' and allied terminology that supposedly have bearing on the moral status of the embryo-foetus. Some of those papers have been referred to in the preceding text.

The Creation is regarded by some as a mere matter of chance, while many of us see the triune Godhead as its originator. As reflected in the early biblical stories, God the Spirit is present as creation takes place, continuing to permeate living matter, so to imbue it with the life-giving force.[31] In the Incarnational narratives, the Spirit overshadows Mary, with the terse reference to that thing conceived within her as being holy, reflective of the Word made flesh (Jn1:14) which 'tabernacles' among us.[32] Life, within that creational context, is seen as sanctified and rendered holy through the Spirit. It follows that the nascent individual from its moment of conception (as I have stressed), becomes a holy, valued being through which its moral status is underpinned and buttressed: therein lies the theological ontology of each newly conceived human individual.

The embryo-foetus is a human being; the *one who is*, thereby encompassing all that pertains to it at whatever time, and in whatever state or condition. It is thus provided with a true anthropological and ontological basis, an individual born into a caring and loving relationship, expectant of its right *to be the one that is* within the tree of human life. And that recognition, horizontally, depends critically and essentially on the continuity of personal relationships and love within its familial context. But that individual, newly conceived, also bears a continuity and relationship, vertically, through the ongoing evolutionary history of which it becomes another link in the reproductive chain. Later, through baptism the individual may be drawn into the death and life of Christ, thereby becoming a recipient of divine grace – or 'divinised'. Moreover, this valence associates that individual with divine Being, and the assertion 'I am that which is'.[33]

We are not just human beings, but humans *becoming*. Becoming is the slow, gradual process of entering that matrix of intertwined communal and familial continuity and relationship which permits that individual to grow and to transcend, as far as createdness allows, one's created physical constraints. But in life, we are cut down, as the late Colin Gunton observed, with some of life's projects incomplete or never started.[34]

That, as he implied, envisages another aspect of continuity and divine relationship which takes each one of us from this physical universe through death and into the transcendently new, resurrection life of the world to come.

* * *

1. Goodpaster K, 1978. On being morally considerable. *J Philosophy* 75: 308-25.
2. Warnock GJ, 1971. *The Object of Morality.* London: Methuen.
3. Warnock M, 1984. Report of the Committee of Inquiry into Human Fertilisation and Embryology. London: Her Majesty's Stationery Office.
4. The Abortion Act, 1967, provides legal criteria under which the uterine contents may be evacuated, under normal circumstances restricted to 24 weeks. But technically, there are no time constraints on such actions, because of other unforeseen eventualities that might arise at any time throughout pregnancy and then probably requiring operative procedures.
5. Law Report, 2015. Mother not criminally liable for alcohol damage to foetus. Court of Appeal, *The Times* newspaper (London), Jan 14th 55.
6. The adjectival meanings of 'individual' have wider usage (where the noun could be inappropriate), in such phrases as individual animal (or cat, tiger), child, adult, or single member of class, society, species or thing within a group of objects.
7. This is the first cell of the embryo-foetus, formed from fusion of ovum and sperm, and containing a completely new and unique human genome (48 chromosomes).
8. Olson E, 1997. *The Human Animal.* Oxford: Oxford University Press, 73ff.

9. Burgess J, 2010. Could a zygote be a human being? *Bioethics* 24: 61-70.

10. Rorty A, 1990. Persons & Personae. In Gill C (ed), *The Person and the Human Mind.* Oxford: Oxford University Press 21: and Morton A, Why there is no Concept of a Person?, idem 39.

11. Sapontis S, 1981. A critique of personhood, *Ethics* 91: 607-618.

12. Teichman J, 1985. The definition of person. *Philosophy* 60: 175-185.

13. Shakespeare T, 1998. Choices and rights: eugenics, genetics and disability equality. *Disability & Society* 13: 665-681: also Kerr A & Cunningham-Burley S, 2009. On ambivalence and risk: reflexive modernity and the new human genetics. *Sociology* 34: 283-304.

14. It is interesting to recall the various manipulations (if not contortions) employed in defining the *in utero* origins of individuals, many relying on:

    1: some kind of brain function (Gertler G, 1985-6, Brain birth: a proposal for defining when a foetus is entitled to human life status, *S Calif Law Rev* 59: 1061-1078: Shea M, 1985, Embryonic life and human life, *J Med Ethics*, 11: 205-209);

    2: an 'active brain' (Goldenring J, 1985, The brain-life theory: towards a consistent biological definition of humanness', *J Med Ethics* 11: 198-204: Cornwell J, 1986-7, The concept of brain life: shifting the abortion standard without imposing religious values, *Duquesne Law Rev* 25: 471-479), or;

    3: neocortical activity sufficient to support those capacities of sentience upon which a concept of personhood might be predicated (Flower M, 1985, *J Med Philosoph*, 10: 65-75: Fletcher JJ, 1974. Four indicators of humanhood: The enquiry matures. *Hastings Center Report* 4: 4-7.

    Are there, we might wonder, valid empirical data underpinning each of these varied claims?

15. Wyatt J, 2001. Medical paternalism. *J Med Ethics* 27 [Suppl II],

15-20: Chervenak F & McCullough L, 1996. The fetus as a patient: an essential ethical concept for maternal-fetal medicine. *J Maternal-Fetal Med* 5: 115-119.

16. Williams C, 2005. Framing the fetus in medical work: rituals and practices. *Soc Sci Med* 60: 2085-2095.

17. Glover V & Fisk N, 1999. Fetal pain: implications for research and practice. *Brit J Obs Gynae* 106: 881-886.

18. Hepper P, 1993. Prenatal and perinatal behaviour. *J Reproduct Inf Psychol* 11: 127-128.

19. Moon C *et al*, 1993. Two-day-olds prefer their native language. *Inf Behav Devel* 16: 495-500; Mampe B *et al*, 2009. Newborns' cry melody is shaped by their native language. *Curr Biol* 19: 1-4; DeCasper A & Spence M, 1986. Prenatal maternal speech influences newborns' perception of speech sounds. *Inf Behav Develop* 9: 133-150.

20. On the control of pregnant women and the increasingly seen importance of the foetus, see: Markens S, Browner C and Press N, 1997. Feeding the fetus: on interrogating the notion of maternal-fetal conflict. *Fem Stud*, 23: 351-372.

21. Berrebi C, Ostwald J, 2014. Terrorism and fertility: evidence for a causal influence of terrorism on fertility. *Oxf Economic Papers*, doi:10.1093/oep/gpv042.

22. The influence of maternal stress through pregnancy should not be underestimated, nor its influences on the developing foetal brain: Heuser I & Lammers C-H, 2003. Stress and the brain. *Neurobiol Aging*, 24: S69-76.

23. Heijmans B, Tobi E, Stein A, Putter H, Blauw G, Susser E, 2008. Persistent epigenetic differences associated with prenatal exposure to famine in humans. *PNAS [USA]* 105: 17046-17049.

24. Bygren L *et al*, 2001. Longevity determined by paternal ancestors' nutrition during their slow growth period. *Acta Biotheoret* 49: 53-59: Kaati *et al*, 2002. Cardiovascular and diabetes mortality determined by nutrition during parents'

and grandparents' slow growth period. *Eur J Hum Genet* 10: 682-688: Lumey L *et al*, 2007. Cohort profile: the Dutch Hunger Winter families study. *Int J Epidemiol* 36: 1196-1204.

25. Ford N, 1988. *When did I begin?* Cambridge: Cambridge University Press.

26. James W, 1988. Anomalous X chromosome inactivation: the link between female zygotes, monozygotic twinning, and neural tube defects? *J Med Genet* 25: 213-216.

27. Ingerslev J *et al*, 1989. Female haemophilia A in a family with seeming bidirectional lyonisation tendency: abnormal premature X-chromosome inactivation? *Clin Genet* 35: 41-8: Jorgensen A *et al*, 1992. Different patterns of X inactivation in MZ twins discordant for red-green color-vision deficiency. *Am J Hum Genet* 51: 291-298.

28. Kaplowitz P *et al, 1991.* Monozygotic twins discordant for Ullrich-Turner syndrome. *Am J Med Genet* 41: 78-82.

29. Ullrich-Turner syndrome is produced in a single, X chromosome-bearing individual (XO) comprising, phenotypically, an infertile female of short-stature, with a webbed neck and abnormally angulated arms. Discordance means only one of MZ twins having the disorder.

30. Schmidt R *et al*, 1976. Monozygotic twins discordant for sex. *J Med Genet* 13: 64-79.

31. Gen 1:2. The Hebrew verb (mrakhephet) variously translated suggests hovered, brooded or moved over.

32. Brown D & Loades A (eds), 1996. The Divine Poet. In: *Christ: The Sacramental Word*. London, SPCK 1-25.

33. Exod 3: 14.

34. Gunton C, 2002. *The Christian Faith*. Oxford: Blackwell 57ff.

Part III

Disablement

Chapter 8

On Being Disabled,
Dysfunctional, & Disfigured

One striking aspect of many publications on humankind is the often scant reference, if any, to the embryo-foetus, neonates, children, or to those – the subjects of this chapter – who are disabled, dysfunctional or disfigured.

It may be that being a disabled person rarely engages the thought-processes, for example, of philosophers, anthropologists, or even theologians. That, observed from my previous background of academically based clinical practice, seems quite extraordinary.

But humans can be very 'two-faced' wittingly or unwittingly. We can pursue, along parallel non-interacting trajectories separate thought processes, actions and their consequences, and yet be entirely immune to their impact on others. Think of photographed Nazi leaders surrounded by a loving adoring family while, during office hours, perpetrating the worst atrocities against children and adults known to mankind. Or members of the UN Security Council who, while upholding mandate to seek world peace, sell with their other hands munitions to warring factions worldwide.

Disability is categorised by WHO as physical, cognitive, affective, developmental.[1]

Disability is an umbrella term, covering impairments, activity limitations, and participatory restrictions. An **impairment** is a problem in body function or structure: an **activity limitation** is a difficulty encountered in executing a task or action: while a **participation restriction** is a problem involving life situations. Thus disability is a complex phenomenon, reflecting an interaction between features of persons' bodies and minds, and features of the society that they inhabit.

Ten million UK people are disabled. Seven million are of working age but only 50% work, 23% having no qualifications (compared with 9% non-disabled). Just over ¾ million disabled are children and approximately 17% citizens are disabled from birth.[2] In terms of type dysfunctions, 0.6 million are totally or partially blind; nine million have auditory difficulties; one million suffer mental impairments, while four million have arthritic/musculo-skeletal disorders. Clearly, disablement is widespread, involves many of those of working age, and represents a major loss of economic power for country and victim.

Although envisioning disability from the narrower perspectives of function and economy fails to see disability resulting either from chance, or the imposed injuries of other agencies, human rather than geophysical. Furthermore, although perhaps casually not so regarded, disability, dysfunction and disfigurement of whatever origins are very much a reflection of evil. Thus the question arises why any type of evil should have to occur in a world created by a supposedly all-benevolent, all-powerful God.

This chapter seeks to evaluate disability from these wider horizons.

Embracing the Wider Picture – Is Our Approach Sufficiently Inclusive?

The data, given above as cold facts, reflect a spectrum ranging from the very severely disabled where round-the-clock care is necessary to those with surmountable problems: or, as someone remarked, '...like dealing with a very unruly little terrier!' But viewing those data as a *spectrum* of disabilities further emphasises there is no cut-off point between so-called 'normality' and 'abnormality'. Yet our attitudes foster them-and-us divides. In that we are all heirs to the inbuilt errors, frailty and corruption of the flesh, we are all vulnerable. A broken bone focuses our vulnerability: no one is exempt. To push that point even harder, we must realise that our bodies progressively deteriorate with age assuming liability for disablement, through arthritis, osteoporosis, dementing brain degeneration, and malignancy with the aftermath of its therapeutic interventions.

And in *Bodies* the authoress reminds us through one word (although we should be aware of many others), that there is no necessity for any profound disablement to be publicly manifest.[3] The word is 'impotence', and it begins to make it appearance in 40-year-old men. Thus, relatively early in life there arises a sense of failure, even the coming realisation perhaps of a broken marriage with all its attendant, deeper psychological outcomes.

From that emerges the clear message that disability should be seen by society as an inclusive state of affairs. Yet that is a message hardly recognised universally, let alone accepted. Society continues to walk by on the other side.

The young rugby international Matt Hampson, rendered paraplegic when his neck was broken beneath a collapsing (rugby) scrum recalls – sometime later – losing respiratory control in his wheelchair due to a flat battery. As a replacement was awaited, his carer initiated manual ventilation of his lungs with a bag. 'Meanwhile', he relates, 'it was business as usual on Broad Street [Birmingham, UK]...Nobody offered to lend a hand...Nobody offered to call an ambulance... I was stranded on one of Britain's busiest streets'.[4]

In reading that account, my mind wandered back to another stark, albeit imaginary, portrayal in the same city:

When Jesus came to Birmingham they simply passed Him by...They only just passed down the street, and left Him in the rain...Still Jesus cried, "Forgive them for they know not what they do", And still it rained the wintry rain that drenched him through and through...and Jesus crouched against a wall and cried for Calvary.[5]

In addition to the obvious *indifference* (the title of Studdert Kennedy's poem) by the majority towards disability and of recognising the need for and opportunity to give aid as and where required, is our *impatience* with it. In our mad rushing 'to-get-there-on-time', we can be exasperated when impeded by the slow moving wheel-chair or halting progress of the individual with a stick.

Hurrying may not simply indicate indifferent callousness, but rather pervasive and possibly urgent, need-to-be-met obligations: thus one's conscious-awareness is totally oriented towards these latter needs. This was demonstrated by selected Princeton seminary students told to 'hurry' to give a talk on either ministry or the Good Samaritan, and on their way being forced to pass an 'ailing' actor.[6] Only other seminarians, told they had plenty of time before speaking, stopped to offer help and comfort. Even the volunteer with a collecting box serves to reinforce, negatively, that divide – even though we might momentarily yield and embarrassedly fumble for a coin. Yet in acting for the cause, a hint of paternalism seems to overshadow our gesture towards those unfortunate to be 'there' and in need of financially (charitable) supported care. Moreover, giving is impersonal, importantly demanding of us no further commitments or responsibilities thereafter.

The academic John Hull has articulated a particularly valid point of view, engendered by his blindness.[7] His condition emphasises that perceptions of the world are heavily dependent on our sensory intakes. There are neither universal nor valid blueprints delineating the world of imagined normality. Remember, it is our brains which create the verisimilitude (to use John Polkinghorne's favoured word) or 'virtual image' in which rests the reality of the substantive world external to our senses.

There was recent outcry about the supposed immorality of a deaf couple wishing to conceive a deaf child.[8] Despite the moral ambiguity, it is not novel. Deafness increased after the introduction of American Sign Language. Why? Because deaf folk, having now come into a communal relationship to learn the language, were improving if not enjoying their newly-found social capacities. Resulting from acquaintances freshly acquired, some became married thereby helping to spread one predisposing gene for the impairment. Note: attempts to ameliorate any defect may well throw up unexpected consequences. To deaf parents, however, a deaf child is not 'abnormal'. And, to what extent in today's society should we consider the life of a deaf person to be either impoverished or undermined? We must certainly learn not to impose our presupposed criteria of 'normality' onto those who do not warrant such incursions: how can we possibly know?

My niece's daughter with Down's syndrome is a very self-sufficient (almost confident!) 15-year-old enjoying life at her level of achievement. Why should 'normal' people have demanded abortion? Down's syndrome is the commonest reason for abortion. Furthermore, on what grounds could this young girl's performance be judged to be substandard? The case exemplifies the moral impropriety of demanding solutions for 'imperfections' considered, from others' detached levels of 'normality' as repulsive. But who, dare we ask, assays our own varying levels of achievement (or, for that matter, 'normality') – performer or observer?

Therefore, there are only pluralistic outlooks based on individualised worldviews, heavily biased by personal upbringing, locale, and pre-suppositions. There is no view from 'nowhere', and certainly no 'perfect', untarnished generic viewpoint which we, the so-called able-bodied, inhabit over and against that of the disabled. Thus, for those capable of doing so, the worldview from a disabled perspective should be articulated and absorbed, in counteracting the arbitrary divide between 'unnatural' and 'natural' worlds.

These examples strikingly attest to the moral impropriety of demanding a solution which is neither ideal nor indeed warranted from our own detached levels of achievement. The key word here should be 'sufferability' which, as GJ Warnock reminded us in the preceding chapter, is our universal predicament as human beings.

It's All About Our Attitudes

In biblically based human anthropology (Genesis), humankind is fashioned in the 'image of God'. That assertion seems to carry with it the implication that nothing more can, or need be said. I often wonder what response would emerge from those offering that declamatory edict with the sight of a brain-damaged child in a wheelchair as an inclusive member of the divine club. It is all very well to articulate this well-worn 'image' analogy, as if it were an *a priori* assertion. Nevertheless, I take comfort from James Barr's much needed note of caution.[9] In comparing related Hebrew and Aramaic words for 'imagery', his non-committal response is that the phrase may mean very little, particularly in being greatly overshadowed by the far sterner command in Deuteronomy (4:16, 23, 25) not to render any G/god-resembling iconography.

We may be here as God's vice-regents, to think and use our freewill, or perhaps to live in relationship with God and his fellow creatures, developing our God-given capacities to the full, being responsible for the earth, and living out wholesome neighbourly relationships. My point is that such a command has hardly fostered an Edenic paradise on earth. Man's inhumanity to man is there for all to behold: the past and continuing depravity to others within the human race is simply staggering.

Attitudes in society determine and condition responses to others. Western society is subject to secularist precepts, against any hope of immortal existence. From that perspective, our existence is seen as utterly pointless and having but momentary relevance within an evanescent timescale. In this inexplicable

God-less universe, true personhood in neo-Darwinian reductionist dogma is science-based. Thus, in Crick's recent public discovery, we are 'nothing more than our neurons', and so on.[10] Yet surely scientific reductionism, however nuanced, cannot assume a pre-eminent meta-narrative over all other explanations, as Mary Midgley rightly asserts.[11] Moreover, the damage to society which such reductive claims exact should not be ignored. Since secularism eschews all notions of immortality, it is not surprising that people can no longer be regarded as truly one, absolute, dignified, or irreplaceable but simply collectivised masses, or worthless individuals.

First, we no longer inhabit neighbourhoods, from which the cognates 'neighbour' and 'being neighbourly' derive, but 'communities' which lack the corresponding grammatical predicates. People inhabit, so we are informed, vague but 'tightly-knit communities' – a phrase beloved of media reportage. Of a local disaster, it may be reported that the post or milk had been piling up for days around the victim's front door, but 'no-one wanted (or, more truthfully, could be bothered) to interfere'. On other occasions of grief, communities 'come to terms' with unexpected onslaughts, whether caused through ignorance or incompetence, and somehow need 'counselling' as antidote, or 'compensation' – the promised hope of financial reward.

Second, the cult of 'individualism' is foremost in today's social environment. And individualism is the child of 'personal autonomy', the all-pervading gospel that we do things alone without restraint from others, thus to pursue our own relentless furrow, not as part of society nor even of neighbourhood, but apart from it. The procurement of individualised autonomy is potentiated by the mobile phone, laptop, iPad, personal stereo-player, laws forbidding physical contact, indifference to other people. Also, the inevitable wearing of designer 'shades' which obviate the eye contact crucial to fiducial encounter with the other. Wearing the largest, darkest shades, apparently, is the ultimate emblem heralding celebrity status – even though the absurdity of such posturing seems to evade its own recognition! At both extremes, the implied relationship is of 'I-he/she' or even 'I-it', and certainly not 'I-thou' that encourages inter-personal relations, or 'mutuality', as Martin Buber would have put it.[12]

And the 'Professionals' Aren't Much Better, Either

The Equality & Human Rights Commission (EHRC) has reported its surveys of abuses towards the disabled in UK society.[13] People with impairments continue to experience widespread, unremitting assaults of a verbal, physical, sexual or violent nature, resulting in deaths of some victims – either from cold-blooded murder, or suicide by those distraught from unrelieved, targeted abusiveness. A newcomer to the aramentarium of abuse-perpetrators is cyber-bullying, of which 'trolling' is a particularly vicious, and lethal, instrument: we only need to remember how many school children have committed suicide from this influence alone. Slanderous or other degrading remarks about recently deceased victims are also published via internet chat-facilities, bringing additional pain to bereaved parents or relatives. In addition to personal onslaughts, other distressing forms of anti-social behaviour such as damage to property, motor vehicles or personal belongings inculcate uncertainty, fear and vulnerability, which seriously interfere with victims' lives, undermining confidence and robbing them of any pleasure in their existence.

The Commission also found that the mentally-infirm are particularly at risk, resulting in gross underreporting of incidents, unsympathetic hearings from the authorities, and a dismissive attitude towards prosecution since these people are regarded as poor, uncertain witnesses. Fiona Pilkington, a British lady with learning difficulties, together with her daughter Francecca, who had severe cognitive problems, committed joint suicide. That was the culmination of recurrent, unremitting abuse by local children, despite her phoning the police on 33 occasions.[14] Did the local police regard her pleas as the rantings of a 'nut-case' – thus to eschew her *cri-de-coeur*?

Are all the 'lessons' we constantly hear about *really* being learned, and will they apply across the board to all other authorities (whatever their specific 'professional' remit) in light of index cases? And if these lessons are not being (re)learned, why not?

Were 'lessons' (re)learned and universally applied following the Victoria Climbié case: and if so, why did a repetitive series of baby killings follow? The evidence, persistently, says that professional standards generally are neither affected, nor raised to the highest standards, following these tragic occurrences.

And why is it that various types of professional bodies are having to learn on the job? Should not all principles of good practice be instilled during training, applied rigorously by individual practitioners throughout their subsequent careers, and passed on within chains of command and instruction to less competent juniors? Clearly, this is still not the case.

The Commission also notes the indirect reporting of some incidents to third parties, such as care staff, whose roles are uncertain and have yet to be formally ascertained. Moreover, this habit is cause for concern because statutory agency personnel have been shown to be responsible for abuse and assaults, suggesting that reports are unlikely to reach the police.[15] Examples of abuse by care workers and hospital staff, especially in wards or homes for geriatric or mentally ill patients, have been central to recent newspaper reportage or exposures by undercover agents. That so-called vocational health workers are happy to indulge in such terrible happenings clearly is reflective of the general attitudes within society towards the infirm and helpless. These occasions point directly to distancing, repudiation, and not infrequently aggressive attitudes and posturing towards those regarded as worthless, useless, or racially abhorrent. Moreover, that harmless but nevertheless vulnerable babies and toddlers have been subjected repeatedly to the most sickening, repetitive violence resulting in their deaths and often in the most squalid of circumstances (Victoria Climbié, through 'Baby P' to Ryan Lovell-Hancox) is demonstrative of similar attitudes of repudiation within society.[16]

The treatment of humanity as an expendable, commercial, social, political, sexual or even 'fun' object is what the Orthodox theologian John Zizioulas terms 'incapacity'.[17] With such inbuilt themes abroad throughout society, it is not surprising that the disabled, defenceless and weak are herded off, and given a raw deal. The tardiness in the construction of ramps to facilitate access to major buildings is one outstanding case. The climbing of endless staircases to get out of the London underground train system is another analogous, wearisome hazard. In addition to anthropologists and philosophers, it is evident that planners and politicians hardly think in constructive, inclusive ways for the benefit of the disabled who undoubtedly should be automatically thought of as rightful and equal members of society. There is little encouragement from what I have outlined to sustain the idea that we are made in God's image or inhabit an ideal paradise on earth. We need a far more robust theological account of disability to counter the prevalent approaches and attitudes ingrained throughout society.

Fighting Back: Disability Rights & Action by the 'Able-Disabled'

While much disability entails dependency, passivity and waiting, thereby reflecting a Christ-like dispositional 'passion', more recent legal and group action has firmly promoted the cause of the 'able-disabled' and made their voices heard throughout a society which prefers them remaining hidden.[18] Happily, we no longer have court jesters, or see 'monsters' and 'freaks' exhibited for gain or amused curiosity in fairgrounds, shows or carnivals. We should take pride in the accomplishments of the able-disabled within the literary, visual and performing arts, academe, politics, and the sporting world. Here we see an extraordinarily vibrant culture which brings to the fore the clear message that these people are human, an effective part of society, entitled to an equitable and inclusive future, and deserving of dignity equal to 'normal' people. Indeed, we should acknowledge, with Professor Colin Barnes, the long history of wide-ranging exemplary accomplishment effected by the disabled.[19]

But is that enough? Unfortunately not. We continue to stage Paralympics separate from the Olympics proper which, from Christian perspectives, fails the Pauline test: 'You are sons of God through faith in Christ Jesus...for [in baptism]...there is neither Jew nor Greek, slave or free, male and female...for all are one in Jesus Christ' (from Gal 3. 26-8). To live the *ecclesial life* in baptismal solidarity requires a paradigm shift disruptive of assumed personal and socio-political 'norms'. Post-Enlightenment concepts of fully active, autonomous subjects are false and inimical to gospel teaching: we all wait in dependency. Thus, for every broad category – the physically incapable, the mentally handicapped, the able-disabled – there is need for grace enacted through love, gesture and togetherness, and not mere superficial, inconsequential chit-chat about the weak and poor. That achievement requires radically reframed attitudes modelled on Jesus' involvement with the downcast and his promise of an eternal life free from bodily stigma. That is the true meaning of the ethical stance of The Kingdom that Jesus came on earth to inaugurate.

Personal disablement extends far beyond the walking stick or wheelchair. We still have a long way to go.

Why Do Humans Inflict So Much Evil on Other Human Beings?

We cannot quantitate all the good and evil in the world and decide the balance. Overall, most people consider that evil exceeds good although no sound empirical reasons affirm that view: evil is seemingly so pervasive over any good. That, I guess, does not seem to be a bad assessment, given how evil people can be!

What factors make people behave badly towards each other? It is evident that persistent offenders may have frontal lobe brain abnormalities. Antisocial personality disorder (ASPD) comprises 'sociopaths' with mild, and 'psychopaths' with severer, behavioural perturbations. Brain scans of murderers reveal reduced glucose metabolism, and structural changes of pre-frontal cortex and their connections with other important cortical and subcortical areas, especially those modulating emotional responses.[20] Moreover, trauma sustained before 16 months of age can result in non-empathic disruptive social behaviour, insensitivity to punishment, failure to respond to corrective behavioural training, and lack of forward planning.[21]

This is a complex issue involving neuropathology as much as forensic psychiatry. In further evaluating this kind of misbehaviour in order to understand the minds of deviants, we can usefully compare psychopaths with autistic people who also lack a certain empathic stance towards others. But why is it that autistic people do not behave with 'thuggish' intent? We should first note that a lack of empathy – the failure to comprehend others' interests or needs in a non-sympathetic manner – cannot stand as the only explanatory criterion of the psychopath's behavioural defect.[22,23] This seems to be because psychopaths can appear to be very social individuals and thus presumably have functioning mirror neurones as well as a Theory of Mind.

However, despite lack of affective engagement, autistic persons can acquire moral reasoning by intellectual effort alone. Further insights emerge from higher functioning autistics who develop strategies for ameliorating their predisposition. Temple Grandin's life history, outlined in Oliver Sacks *An Anthropologist on Mars*, reveals that she built up a vast library of memories –

'videos' which she repeatedly replayed in her mind to try to know how best to react in different situations.[24] Grandin obtained some semblance of moral reasoning and agency by working very hard at the problem, through gaining a cognitive purchase. This methodology, as Sacks rightly claims, is fundamentally and strictly logical.

Conscience provides reasons that relate our actions to autonomous agency: without proper reasons for action, we should be worried. Reason 'ought' to constrain impaired evaluative judgments. And this is the crux of the difference between the ASPD 'psychopath' and autistic subjects. Because the vital receptiveness to constraints of reason are intact despite lacking empathy, autistic subjects perceive the moral imperative, so that the concerns of others are reason-giving for actionable responses, even if the latter are inappropriate.

Psychopaths, at times, display lack of concern for others' interests, reduced capacities to understand their actions, a failure to review possible available options and thus to act impulsively. This results in sudden uncontrolled outbursts, disorganised and self-destructive action, reflecting an underlying impairment of moral and prudential judgement due to an inadequate conception of proper and appropriate behavioural ends.[25]

Kennett is right to distinguish between the acquisition of moral agency, and the requirements necessary for its proper delivery. Although empathy has its role in moulding the agent into a sensitive responder and thoughtful, caring operative, autism reveals that a moral self and attempted moral agency may be established without it. Conversely, psychopathy reveals that reasons to act are fundamental to right moral agency. 'Only individuals who are capable of being moved directly *by the thought that some consideration constitutes a reason for action* can be[*come*] conscientious moral agents'.[26]

Encountering Personal Tragedy within the Context of Evil: Is there an Answer?

So far, I have outlined the disability curriculum, dismissive societal attitudes towards it, the emergent voice of the 'able-disabled', and those deviant behavioural dispositions causing agential dysfunction and disabling hurt towards victims. That account severely exposes as erroneous the prevailing belief that disablement is the obverse coinage of capability and normality. On the contrary, we are all vulnerable, and no one is exempt. Thus we should peremptorily be reminded that disability is neither a 'them-and-us' phenomenon, nor indicative of a bimodal population.

Secondly, we need to be clear that disability and its entailment is not simply something that happens, whether through natural causes, happenstance or deviant behaviour. We ourselves can be very effective disablement-agents causing injury, dysfunction or assaults on the dignity of others, through our own calculated and deliberate personal acts, omissions, meanness, or from the sheer pleasure and desire of causing harm and discomfort.

But while certain kinds of insult and damage may be surmountable, others are inimical to, indeed totally destructive of, continued flourishing (good-making) in peoples' lives. Such misfortunes come within the scope of 'horrendous evils' envisaged by Marilyn McCord Adams, lately Regius Professor of Divinity, Christ Church, Oxford.[27] These are the individuals who in their deepest misery and unconscionable agony cry out: 'Why this?' 'Why me?' 'What have I done?' 'How could God allow this?'

What kind of God, then, allows such crushing defeats of peoples' lives? We should be under no illusions about the horrific, life-destroying natures of the calamities to which human nature is heir. In her approach, McCord Adams seeks to circumvent the abstract, philosophical intractability of non-resolution: the conflict between God, seen as omniscient and omnipotent, and the obvious fact that evils prevail. Within this quasi-moral framework which she sees both to dominate and obscure the problem of evil, God is reduced to a kind of reprobate, mega-monster who, having been dragged to the dock, is subject to stiff forensic interrogation to answer for his immoral misdemeanours and failings: 'Where were you, God, in my greatest need? and 'Why do you allow such things to happen?'

That a purely moral calculus can be employed as explanation (and even to Divine complicity) may be questioned in that both perpetrator and victim may be entirely innocent of blame.[28] Furthermore that calculus only provides partial answers and frustrates any real meaning. Posed thus, an unresolvable contest is erected between 'that than which no greater being is conceivable' and earthbound creation subject to geophysical, biological and behavioural miseries, terror, calamity and utmost depravity. And since we do not know – nor even come anywhere near understanding – their real meaning, can we really judge why the Godhead allows these horrors? Moreover, if we collapse the divide between God and humanity, how then can we guarantee that He has the power to overcome such evil? And since there can be no answer, *on those quasi-moralistic terms*, the final question rebounds: 'Why, then, do you believe in this so-called God?' – a proposition which for many, is unanswerable and may even erode belief.

Nowhere in the Bible, McCord Adams claims, is God's portrayal exemplified in such simplistic, morally exclusive terms. Adams replaces this conventional view with an 'expanded theism' with which she can do business, a more robust theism embracing traditional approaches to religious understandings and beliefs.[29] Adams seeks to reconfigure Godhead away from purely moralistic and somewhat artificial approaches. In her remodelling, she analyses the codes of honour/shame and purity/defilement.

God as Patron: Exploring The Honour/Shame: Purity/ Defilement Codes

First, Adams employs the honour/shame code that portrays the God-client relationship (Covenants) enacted through the patriarchs, prophets, John, Jesus.[30] We, the chosen, owe obeisance to our divine underwriter for the benefits bestowed. The honour/shame code, in recognising our personal worth and dignity despite our carnal frailty, is the mechanism whereby the Patron's worthiness is expressed.[31] So the narrow moralistic denunciation of God is overcome, and purely materialistic interests in pain/pleasure, disability/normality are sidestepped. And in the final Judgement, God will be vindicated, and be seen to have been consistently steadfast in word and promise.

Next, the purity/defilement code re-establishes God's holiness, set against humankind as unclean, defect-ridden, and disease-prone.[32] The code brings to focus the sheer incommensurability of Godhead as holy, sacred, pure and perfect, and the basal uncleanness (re Levitical code) of animals and humankind (their excreta, discharges and suppurations), their defective bodies (limbless, blind, deaf, diseased, brain-damaged), and the infectious communicability of disease. On this basis is starkly portrayed the holiness/defilement antithesis. The code identifies bodily defilement as the canopy under which so much of humankind's disablement lies, and inimical to being one with Divinity. Mankind's history is a testament to monumental incompetence, thus arguing against any possibility of a 'bottom-up' free-will approach in bridging the gap to secure (Edenic) perfection in this life.

Divine power, rather than human endeavour or piety, histori-cally re-orientates humankind's 'grids of activity', through the guiding providence of God-as-Spirit. Employing the analogy of mother-baby posited by Julian of Norwich, Adams sees the devel-opment of God's earthly children in need of hospitable frame-works, analogously and progressively modelled by parent, home, and family.[33] This is 'agency-developing' and 'agency-forming' (or 'becoming' as John MacQuarrie would have put it: see previous chapter) until coming-of-age rationality decides whether Godhead is honoured or rejected.

Nevertheless, having resolutely re-set the boundaries between God's purity and humankind's defilements, Adams asserts that the key to any fuller understanding of affliction rests with Jesus, God-incarnate.[34]

First, through God-the-Father's love, we are confident that He will guarantee to humanity a life of great good, as exemplified through evolutionary time in the emergence of human consciousness – thereby equipping us to perceive and approach him in reverence and awe. Incarnation reflects the divine love for creation and humanity, while the cross incorporates evil and all its perpetrators into God's universal redemptive plan: depravity is sanctified.

That is, God-as-human (the incarnate Jesus, or God-as-Son) now becomes vulnerable, experiencing at first hand the exigencies of human existence, enduring life's frustrations and frailties, seeing evil wreck individual lives, being subjected to the double-faced outcomes of human behaviour in disownment and desertion, and undergoing complete dereliction in abandonment by the Father. And all that culminates in the ritualised curse of being taunted, beaten and slaughtered as a despised criminal. The Godhead (God-as-Father) also suffers, but more like the parent seeing its child under the dentist's drill: pain deeply felt bilaterally, but qualitatively different.[35]

Second, we can be confident that God-as-Father would not permit any form of suffering without having the power to defeat it. Yet God does not usurp our freedom to act. Suffering could be envisaged by means of categories, like Otto's categories (of fear, trembling, shame) and as another means of encountering the divine through which all affliction and stigmas are given positive meaning, if not partially overcome – symbolically and objectively.[36]

Frances Young emphasises the Cross as the occasion of universal redemption, judgement on all perpetrators of evil, and forgiveness to the believer – but in its obverse, the glory and triumph.[37] I take the Johannine use of the Greek perfect tense 'tetélestai' (Jn 19:30) to imply far more than a simple narrative (aorist) action ('It is finished', and thus often incorrectly translated to imply both the initiation and completion of the Kingdom within earthly time). Surely, as a (purposely written) perfect, it signifies *continuity of accomplishment* as a progressively extending and expanding Kingdom, following its inauguration by Jesus (Lk 4:18-21; 3:4-6: Isa 61:1-3; 40:3-5).

Resurrection heralds the completed redemption of corrupt society and groaning creation, repair of personal damage, and transformation of earthly into spiritual bodies – free of imperfection, disability or disfigurement – and where, for evermore, an everlasting home in Godhead is ultimate promise.[38]

Hope Against Hope

We have travelled extensively in evaluating disability, and perceived (hopefully) its all-pervasive evil. We are all vulnerable to trauma – geophysical, bodily, genetic, as well as life-destroying agency effected through others' cruelty or incompetence – threatening everyone's making-good.[39]

In a universe based only on chance, disability and the evils pertaining to it are mere bits of the hard-luck intrinsic to inhabiting a cold, impersonal, purposeless universe. For the believer, on the other hand, another set of well-versed options is available, albeit neither entirely neutralising any forms of disability, malfunctioning nor, most certainly, the pain of stigma or victimisation: we still inhabit the corruptible.[40]

In forging a sensible theistic approach to the world's evil and the personal havoc it wreaks on making-good; or the notion that Godhead is somehow absent; or that such misfortune is retribution for sinfulness; or that Jesus, crucified, is mere scapegoat to absolve people's wickedness, must all be purged from our thoughts and deliberations.[41]

McCord Adams resists those who anthropomorphise Godhead, blaming it for afflictions from a narrow forensic viewpoint which, she rightly claims, is wrong-footed. Invoking the honour/shame and purity/defilement codes, she restores the metaphysical gap. Cosmic liberation needs God-incarnate: evil becomes defeated by incommensurate Goodness made man-for-man (to paraphrase John Henry Newman).

Young emphasises the duality of the Cross – desertion, desolation, pain, humiliation and God's forsakenness, but also the glorious triumph heralding the ultimate securing of the Kingdom. Jesus, Incarnate, undergoes the very experiences and humiliations endured by humankind throughout its earthly habitation. But in disability, the Incarnate-crucified is there as 'midwife', right at the cutting edge of tragedy when it bites.

Next, disability may be seen as a sign, symbolically pointing to Christ's broken body, but also as a divine calling through which disablement and disfigurement act as beacons, thus again illustrative of Godhead's supreme holiness.[42] Such afflictions or defilements bring us, in passing, to Leviticus 21:15-23. Sarah Melcher rightly distinguishes Levitical behavioural prohibitions deriving from a God-delivered 'strike', discerning with Jennifer Cox that the priest is neither morally nor ritually impure.[43] Cox defines these blemishes as 'enacted parable'.[44] Here, the person – rather than ritual objects used or ritual acts performed – functions as a living sign, being both reflective and exemplary of the otherness, holiness and purity of God.[45] Thus disability, of whatever hue, is *envisaged as divine calling*, analogous to Paul's designation as God-given treasure (2 Cor. 4:7-12) through which the suffering, weak, foolish and the despised will be first-called into the Kingdom (1 Cor. 1:26-31).

Finally, disability conveys a stark reminder of the price paid by Jesus, through his broken wounded body, in coming among us. Although disability is incapable of erasure from biological existence, our inclinations manifest disgust by walking by on the other side. In so doing, we reject God – because the Godhead has identified with our suffering, investing it with holiness in succouring the afflicted.

Disability is thus endowed with deep metaphysical significance and redemptively nullified, symbolically and objectively.

Nevertheless, from uttermost depravity comes hope, a hope in a Power mighty and loving enough to overcome all physical and mental stigmas, thereby affording them a spiritual depth of meaning.[46] Victor Frankl and other writers, demonstrated the power of hope for survival from concentration camp dehumanisation – of *undeviating* hope not forfeitable despite utter desolation: of *undiminished* hope doggedly focussed on returning to incomplete literary composition; a burning love for a distant spouse anticipating imminent widowhood; the passionate reassurance of reunification with a grandchild; or the seasonal cycle of blossom and leaf seen through a chink in the wall of a prison hut.[47]

The thoughts expressed in this chapter can never serve to extinguish the pains of tragedy, disability, disease, the complete wrecking of lives, the mind-numbing disbelief that a life-disrupting disaster has unexpectedly struck – and fallen on *me*; nor the possibility of future occurrence. Those who have experienced such events know this, and as a physician, I know that too. Such events do require tenderness, love, compassion, time-spent, and empathy towards those so mightily burdened, especially when that hurt can never be assuaged during life. To think otherwise, or even to try to magic such evils away is thoroughly mistaken.

Yet in the depths of despair, hope arising from steadfast faith offers succour and deliverance, secured universally and eternally in concrete and personal well-being.[48] That's why the secularists are so ill-equipped to understand the fuller meaning of disability and disablement. Yet those sturdy features of determination, courage and hope have been the testimony of martyrs, of the persecuted, and of individual believers throughout every age.

* * *

1. World Health Organisation, at http://who/int/topics/disabilities/en/
2. London: Office for National Statistics, 2008, and other sources.
3. Orbach S, 2009. *Bodies.* London: Profile Books.
4. Hampson M & Kimmage P, 2011. 'The ventilator stopped...' *The Sunday Times.*
5. Studdert Kennedy G, 1986 (1929). Indifference. In: *The Unutterable Beauty.* Oxford: Mowbray 31.
6. Darley JM & Batson CS, 1973. 'From Jerusalem to Jericho: a study of situational and dispositional variables in helping behavior'. *J Pers Soc Psychol* 2: S100.
7. Hull J, 1990. *Touching the Rock: An Experience of Blindness.* London: SPCK.
8. Deech R & Smajdor A, 2010. *From IVF to Immortality.* Oxford: Oxford University Press 59-60.
9. Barr, J 1968/9. The image of God in the book of Genesis - a study of terminology. *Bulletin John Rylands Univ Library (Manchester)* 11-26.
10. Crick F, 1994. *The Astonishing Hypothesis.* London: Simon & Schuster.
11. Midgley M, 1983. Selfish genes and social Darwinism. *Philosophy* 365-377: Idem, 1992. *Science as Salvation.* London: Routledge 40ff.
12. Buber M, 1966. *I and Thou.* Edinburgh: Clark 15: 33.
13. EHRC, London: 2009. *Promoting the Safety and Security of Disabled People.* EHRC, London: 2011. *Hidden in Plain Sight:*

Inquiry into Disability-Related Harassment.

14. Jardine C, *The Daily Telegraph* 2011. 'Fiona Pilkington: will we hear the next cry for help?'

15. Pistorius M, 2011. *Ghost Boy.* London: Simon & Schuster. For several years this young lad lay in a bed apparently unconscious, although he was actually suffering from a very rare type of 'locked-in syndrome', possibly of infective aetiology. Although completely paralysed, he could nevertheless see, hear and memorise everything that happened to him. Vomit was reintroduced into his feeding tube (p161), and his toothbrush washed with faecal enema fluid (p162): one 'carer' used to take him away from the ward, pull up her uniform, and pleasure herself on his body (p164).

16. Bennett R, 2011. 'Toddler's killers were known as a danger to children'. *The Times* (London), 2011. 'Baby P' was killed ten years after Victoria Climbié, and in the jurisdiction of the same Children's Department, Haringey Borough Council, Greater London.

17. Zizioulas J, 1975. Human capacity and incapacity: a theological exploration of personhood. *Scot J Theol* 28: 401-448.

18. Vanstone W, 1982. *The Stature of Waiting.* London: Dartman, Longman & Todd.

19. Barnes C, 2013. Generating Change: Disability, Culture & Art. *Behinderung und Dritte Welt* 19: 4-13

20. Raine A, Buchsbaum M, LaCasse L, 1997. Brain abnormalities in murderers indicated by positron emission tomography. *Biol Psychiatr* 42: 495-508.

21. Anderson S, Bechara A, Damasio H, Tranel D, Damasio A, 1999. Impairment of social and moral behaviour related to early damage in human prefrontal cortex. *Nat Neurosci* 2: 1032-1037.

22. Kennett J, 2002. Autism, empathy and moral agency. *The*

Philosoph Quart 52: 340-352.

23. Marsh MN, 2003. Empathy: Mirroring Another's Predicament – or More? *Antonianum* 3: 409-430.

24. Sacks O, 1995. *An Anthropologist on Mars.* London: Picador 248.

25. Elliott C, 1992. Diagnosing blame: responsibility and the psychopath. *J Med Philosoph* 17: 200-214: Kennett 2002, 353-6.

26. Kennett 2002, 357: my italics.

27. Adams M M, 2000. *Horrendous Evils and the Goodness of God.* London: Cornell University 27ff: 80ff.

28. For example, a woman stopped her car to assist a man lying on the ground. She failed to apply the brake, so that as she left the vehicle it ran over the man and killed him. Similar tragedies occur on front driveways when toddlers, thought to be in the house, are run over and killed as the car is reversed out of the garage.

29. Adams' rejection is against what she terms the philosophically-based 'restricted standard theism' of Mackie, Plantinga and Hick: reprinted in M.M. Adams & R.M. Adams (eds), 1990. *The Problem of Evil.* Oxford: Oxford University Press.

30. Adams, *Horrendous Evils* 112ff.

31. Adams, *Horrendous Evils* 125ff.

32. Adams, Horrendous Evils 86ff. Also Douglas M, 1966. *Purity & Danger.* London: Routledge.

33. Adams 2000, 104.

34. Adams, *Horrendous Evils* 164ff.

35. Adams, *Horrendous Evils* 174. The parental metaphor is demonstrative of much biblical reference: see Crouch CL, 2010. Genesis I:26-7 as a statement of humanity's divine parentage. *J Theol Stud* 61: 1.

36. Adams 2000, *Horrendous Evils,* 161-4: references here to the 'grip' of the divine beloved (S Weil), or of God-the-Father greeting us post-mortem because of our life-destroying

suffering (Julian of Norwich) – much as we suppose He would give welcome to martyrs for having forsaken their lives for the Kingdom.

37. Young F, 1982. *Can these Dry Bones Live?* London: SCM Press 60-1.
38. Hebblethwaite B, 1996. *The Essence of Christianity.* London: SPCK 189-190.
39. Adams MM, 2006. *Christ and Horrors: The Coherence of Christology.* Cambridge: Cambridge University Press 194ff.
40. Cottingham J, 2005. *The Spiritual Dimension: Religion, Philosophy and Human Value.* Cambridge: Cambridge University Press. I am rather attracted to his (Liebnizian) sentiments that Creation, not being God-like, must therefore be evil and degenerate 27-29.
41. Hughes G, SJ, 2007. *Is God to Blame?* Dublin: Veritas Press 72.
42. Cox JA, 2011. Disability as an Enacted Parable. *J Religion Dis & Health..* 15: 241-253.
43. Melcher S, 1988. Visualising the perfect cult: the priestly rationale for exclusion. In: Eiesland N & Saliers D (eds), *Human Disability and the Service of God.* Nashville: Abingdon Press 55-71.
44. Blindness; withering or breakage of a limb; lameness or hunchback; a broken nose; scabbing, oozing of skin, and an *undescended* testicle [that's how I translate the Hebrew: a 'crushed' testicle, as usually rendered, has no clinically meaningful value whatsoever]: the latter would be noticed on the 8th day at circumcision – hence the reason why a non-observable defect would become public knowledge.
45. Hooker M, 1997. *The Signs of a Prophet.* London: SCM Press.
46. When confronted by any type of adversity, most people do *NOT* commit suicide. As Richard Harries [former Lord Bishop of Oxford] once wrote, in the most abject of circumstances, people still rejoice at the birth of a baby, or on seeing a glorious sunset. Harries R, 1995. Evidence for God's

Love. In: *Questioning Belief*. London: SPCK 14-15, 27.

47. Frankl V, 1985. *Man's Search for Meaning*. New York: Pocket Books 90ff: Gollwitzer H, Kuhn K, Schneider R, 1974. *Dying We Live*. London: Fontana (German text).

48. Adams, *Horrendous Evils*, 177. Individual testimonies (name & address supplied, as it were) are carefully detailed by M. Baughen in his *The One Big Question* (Crusade World Revival, 2010) providing strong evidence of undoubted hope and confidence through steadfast faith that life's sufferings will be reversed in resurrection.

Part IV

Disposal

Chapter 9

The Embryo, Abortion & Infanticide: Biological Thoughts & Ethical Concerns

Abortion and infanticide are procedures destructive of nascent human origins. Their permissibility could be viewed as polarised by secularist or religious opinion. One option is shaped by 'pro-choice' rights of women to control their own bodies. 'Pro-life' groups, upholding 'sacredness of life' values, argue against wanton destruction.

Secondly, philosophical enquiries into abortion and the right-fulness of killing young humans involve controversial defini-tions of a 'proper person', or the application of the 'utilitarian-consequentialist principle' – seeking the greatest happiness and pleasure for the majority. Some philosophers, while agreeing that embryo-foetuses are human, exclude mere species membership from moral consideration, thus allowing abortion and infan-ticide. Similarly, because infants lack true personhood, they are excluded from the 'moral community' and can be killed, their body parts more usefully transplanted elsewhere.

Third, disagreements concern the time of origin of 'persons' (Chapters 6 and 7). *Biologically*, criteria include the appearance of the primitive streak; the latest opportunity for twinning (each occurring ~14 days post-fertilisation); brain development and sentient function, or 'quickening'. *Morally*, 'personhood' and its parallel 'having a life' are widely accepted to reflect capacities exhibited only by competent adults, as noted above, to the disad-vantage of embryos, neonates, children and the mentally handi-capped.

A Christian perspective on these issues needs to be developed to assess the virtue or otherwise of abortion practice and infan-ticide. For example, there is little chance for manoeuvre if the earliest embryo is regarded simply as a meaningless clump of cells to be snuffed out without hint of moral disgust: so also if severely damaged infants are traded as sources of spare-part organs for transplantation surgery.

I offer simplified empirical, biological data whereby some of these claims can be rebutted: Christians need some understanding of the relevant biology (however taxing!). Second, I suggest that the other positions mentioned above can be robustly challenged. Through such challenges, ethical horizons are widened thereby helping us to focus more critically upon the loss of life involved and the practices, both legal and medical which, in part, condition them.

Finally, I should add that this chapter may revisit themes explored in previous chapters in this book (Chapters 6 and 7). But since each chapter can be read as a separate piece, it is better to use material as required, rather than ask the reader to continuously refer back to other parts of the book when necessary.

Legal Background & Provisions

The abortion/infanticide literature is one-sided because statistical data rarely come into view. Conditioned by that strange deficiency, we can hardly evaluate either those involved in, or the recipients of these procedures. Given the data, however, we can ascertain the various reasons why embryos and newborns are destroyed, for whom, and numbers involved.

Under English law, abortion is *still* a crime. Blackstone in 1765 terms the killing of a live foetus a 'misprision', later decreed murder [1839: Offences against the Person Act (S.58)].[1,2] Abortion therefore remains a felony deserving 'penal servitude for life' (S.58, 1861), while aiding abortion entails imprisonment for three years (S.59 (1861). For killing a child while actually being born (other than for saving the mother's life) the sentence is life imprisonment [Infant Life (Preservation) Act, 1929].

Contrary to apparent widespread public misunderstanding,

abortion is *still illegal*. The 1967 Act selectively permits lawful
abortions by registered medical practitioners in authorised
premises under defined circumstances.[3] The Act does not define
such considerations as: 'risk to life', 'injury to the physical or
mental health of the woman', or 'injury' sustainable by pre-
existing children.

Certification only provides categories. Thus we are ignorant
of the reasons causing women to seek abortion, and what risk or
injury to, or for, themselves they possibly foresaw through being
pregnant. The Act specifies women's safe evacuation and is
disinterested in the embryo-foetus, which remains unprotected
in law. The certificates are deemed confidential, neither being
revealed to the police nor subject to peer review. If that obtained,
doctors committing misdemeanours could be disciplined by the
General Medical Council. Indeed, the Act was framed to counter
a previous judgement given against the surgeon involved.[4]

Numerical Aspects of Abortion Practice

Worldwide, more than 42 million abortions are performed
annually – many presumed illegal. This paper is concerned with
data from the UK where up to two million abortions are
procured annually.[5]

Excepting Category E (malformations), all certifications since
2000 (Office for National Statistics: ONS) have virtually
employed Category C – 'that pregnancy continuing beyond 24
weeks would be more harmful to the physical or mental well-
being of the woman'. Category C, acting as a 'catch-all', predominantly involves (50%) young, unmarried women aged 15-24.
Furthermore, 47% of UK abortions are now procured with
medical abortifacients before the ninth week after fertilisation.
Clearly, abortion is the preferred 'contraceptive', despite other
effective preventatives. In total, up to eight million UK abortions
have been performed since 1967.

On an annual basis, about two million abortions are procured across Europe, UK, and North America and Canada.

In England, two doctors supposedly acting independently of each other have to agree that the intended abortion qualifies. Given so many abortions, it is curious that there never seems to have been a single disagreement between them. Moreover, since a pregnancy can now be detected at a very early stage, the provisions of the UK (1967) Act are becoming increasingly more difficult to abide by. Figures published by ONS indicate that some medical practitioners flout the provisions of the Abortion Act and that their performance becomes a thoughtless rubber-stamp exercise. Despite the Act's provisions, we should note the creeping, careless concerns for life and law thereby engendered. The disregard for the law is again referred to (Chapter 10) in my considerations of how legal provisions for end-of-life assisted dying should be framed.

Data for Abnormal Pregnancies & Birth Defects (infanticide)

Category E (S1. (1)(d)) permits legal abortion at *any stage* of the pregnancy if there is a substantial risk that the foetus, if born, would 'suffer sufficient physical or mental disability resulting in serious handicap'. In 2010, there were 2,290 (1%) abortions under this head, 2143 *before* 24 weeks gestation, and 147 *after* 24 weeks. Category E abortions consistently occupy 1% of UK abortions.

The grounds under Category E abortions may be classified briefly as follows:

(a) *Nervous system malformations* (456 *before* 24 weeks, and 66 *post* 24 weeks) that is, 522= 456+66. Similar nomenclature is used below.

(b) *Congenital malformations of other major systems or organs* – including heart, muscle, bones, and urinary system accounting for 1,094 (988+106).

(c) *Chromosomal aberrations* (831=802+29) of which the

majority were for trisomy 21 (Down's syndrome (482),
Edward's syndrome (164), or other anomalies (185));

(d) *Additional factors* (365=353+12) such as influence of
maternal lifestyle (115) although particulars not listed;
problems with foetal growth (16); foetal/maternal blood
group incompatibility (31); family history of hereditary
diseases which are unspecified (181); and reduction of
maternal load from multiple pregnancies (22).

A further notification system (through ONS) is The National
Congenital Anomaly System (NCAS). It lists those babies
surviving pregnancy but delivered with congenital abnormalities. In the UK (2008), 4,254 babies were born with either one or
more congenital abnormality. Brain and neurological abnormalities accounted for 223 registrations; cleft lip and palate 429;
cardiovascular system 1004; and Down's syndrome 290. These
data permit more informed evaluations of abortion and infanticide. We cannot infer how many 'malformed' infants die, or
how.

Foundational Stages in Embryo Formation

Before proceeding with ethical concerns, I revisit the embryo's
early formative stages, extending the simple account given in
Chapter 7. In doing so, I expose varied assertions that assume the
(absolute) permissibility of destroying the early embryo and
young child with impunity.

Processes involved in the creation of embryos are extremely
complex: happily, our knowledge continues to expand rapidly.
My concerns are that much writing about the moral status of the
embryo comes from those not trained in basic science and whose
working knowledge of current empirical science is sometimes, if
not amazingly, insecure. But that may be part of a strategy
intended to rob the early embryo of moral status thus permitting
its destruction with impunity.

However, some contributors to that literature are members of governmental or presidential committees, influencing political and judicial opinion. Therefore, I advance the case that a biological point of view is relevant to these important discussions, and should temper opinions and arguments concerning the rightness of abortion and infanticide.

(*i*) *Ovum, Fertilisation, Syngamy and Zygote:* in this subsection, I consider the ensuing events once the paternal gene set enters the ovum, fully documented by the Carnegie Institute's recent new project 'The Virtual Human Embryo'.[6]

First, considerable activity ensues during the 16-20 hour period following fertilisation (Carnegie Stage 1a). Chemical signals deriving from the sperm release specific enzymes (phospho-kinase C), which actively stimulate the ovum's chromosomes to divide.[7] We should be very clear that the ovum cannot be fertilised until it has been made ready: *it is not fully operative until the actions of the penetrating sperm make it so.* In the presence of the sperm head, the ovum is stimulated to progress beyond a state of 'maturational arrest' that persisted since the female producing that ovum was, herself, in her mother's womb and during which her (maternal) foetal ovaries were being developed (here again, as in the previous chapter concerning the moral status of the embryo, we see the involvement of three generations – grandmother, mother and intended infant). That is a most important fact to be understood. Once this final maturation is achieved (by chromosomal division: meiosis II, or 'MII' in abbreviation: Carnegie Stage 1b), the ovum becomes functional. Then, the two gene sets come together (termed 'syngamy') thus to yield the first embryonic cell, or 'zygote' (Carnegie Stage 1c).

Second, the sperm contributes the all-important 'centriole', a structure that forms the necessary framework for allowing the zygote to undergo its first division, thereby completing the two-cell (blastomere) stage of embryonic development. Again,

without the centriole, brought and deposited by the sperm, the
new individual could never emerge.

Third, if we can take cues from the zebrafish, sperm
penetration reorganises the ovum's interior, thus being responsible for setting down certain plans anticipating the embryo's
future planes of orientation, and shaping the three-dimensional
organisation of the emerging embryo.[8]

From all that, it can be seen that the assertion that during
these very early stages, all the action derives from the ovum, and
is therefore maternally powered, is false. From that, it is inferred
that the embryo is neither an independent, coherent organism,
nor a 'person'. That is clearly not the case.

Indeed, for zygote development, the crucial role played by
the sperm in activating the ovum and initiating the subsequent
train of events must, from now on, be clearly understood.
Viewed biochemically and energetically, fertilisation-activation
is an explosive event, signalling to all parts of the ovum's
cytoplasm that fertilisation has occurred. In effect, the ovum is
'informed' that it is *no longer* simply an ovum – a critical fact
rarely acknowledged, if actually known and understood. Sperm
activation results in recurring waves of calcium ions (a universal
mechanism for inter/intracellular communication) which flood
the egg's interior. This is an enormous task since, compared with
the miniscule sperm's head, the ovum is less than 100
micrometres (μm) diameter (a red blood cell, comparatively, is
7μm diameter).

So, while maternal energy nourishes the nascent embryo
through its early autonomous stages until embedded, there is
much more to the story. Maternal influences stimulate the follicle
from which the mature ovum is released and made ready for
fertilisation. Other maternally-derived stimuli guide the sperm
towards the ovum, by means of the sperm's forward nasal-type
epithelium which 'sniffs' out these signals. Later, once
embedding has taken place, nutrients from the maternal blood

continue to feed the foetus via the placenta (a paternally-derived organ) until delivery. During the post-natal period, maternal milk nourishes the infant until weaning has taken place.

Clearly, maternal influences provide ancillary support lasting one to two years, but they are not the *only* factors operative between fertilisation and zygote formation. Thus, it is a false claim of several writers to dismiss the earliest stages of embryo development because the latter are supported, partially, by pre-existing maternal energy sources, thus to conclude that the embryo is not independent or self-sufficient.[9] Indeed, from the data offered, sperm plays a major role in actively initiating the ensuing developmental cascade.

(ii) From Zygote to Uterine Embedment: many philosophers actually think that the post-zygotic embryo is an inconsequential, nondescript miscellany of blastomeres whose destruction is not morally relevant.[10] Another view is that since early blastomeres, if separated, might themselves grow into new embryos (because blastomeres are embryonic stem cells), the early embryo is not a composite, cohesive organism and could not deemed to be so, until later phases in development.[11] Such assertions might be considered cogent philosophical propositions, but – in lacking biological integrity – they are wrong.

Blastomeres are pluripotential, but that fact, viewed either from moral or philosophical perspectives, is of trivial signifi-cance, especially when linked to another meaningless notion that newly created embryos are mere 'bags of marbles'. Recent studies have clearly demonstrated that cell-cell contact is an absolutely crucial requirement for proper and continued development, as the initial two blastomeres progressively increase in number.

Firstly, blastomeres cultured individually develop haphaz-ardly becoming, over time, membrane cells ('trophectoderm' or amnio-chorionic), rather than concurrently differentiating into embryo cells (the 'inner cell mass') from which the new individual arises.[12] Secondly, the 'polarity' (top and bottom) of

these cells is poorly developed, due to low levels of contributory
cytoskeletal ('scaffold') proteins. Inter-cellular contacts and the
development of cell polarity are of critical importance to the
autonomous embryo as it enlarges from zygote to the implan-
tation stage: those facts need acknowledgement. These are the
means through which co-ordinated gene expression occurs,
guiding blastomeres towards their ultimate fate and polarity,
and to their correct positioning for the necessary, orderly three-
dimensional assembly of the nascent individual.

Thirdly, observations on human blastomeres taken from
2,4,8,16-cell stages demonstrate that the embryonic genome is
functional immediately after zygote cleavage.[13] This is far earlier
than presumed hitherto, indicating that embryonic gene
transcription ('reading the DNA script': see Chapter 3) is
operative immediately following syngamy. From the period of
syngamy to blastocyst formation, just under 2,000 gene-based
active transcripts have been characterised – so far.

As might be expected, 'pluripotency' (the ability to become
changed into any specialised cell type) persists. In fact, despite
the considerable fuss made over the pluripotency of early
blastomeres by bioethicists and philosophers, pluripotency
continues for some time after the blastocyst embeds in the
uterine wall, since it is necessary that all 250 differentiated cell
types identifiable within the (adult) human body must be genetically specified and directed to their rightful positions, as each
subsequent developmental stage in foetal growth and development is brought about.

Contributions of Biology to the Ethics of Individuals, & Abortion

The foregoing affords a detailed current description of the inter-
related events discernible during the earliest moments of embryo
formation. Their sheer complexity is awesome, and should dispel
any residual beliefs that early embryos are entirely incoherent.

That is an important statement in countering those who incorrectly, and in complete ignorance of these metabolic activities, insist otherwise.

The crucial role of the penetrating sperm is central in initiating the subsequent molecular cascade – activating the ovum beyond its embryonically-arrested final division, organising the future three-dimensional embryonic plan, and providing the centriole through which syngamy and the 2-cell blastomere stage actually become possible. Moreover, it should be clear that the 2000 gene-activations (capable of detection to date with currently available techniques) resulting in the blastocyst and its embedding into the uterine wall, detract incisively from the casual, uninformed idea that the embryo is but a tangled miscellany of nondescript cells.

Thus the American philosopher J. McMahon is happy to declare:

> There is no deep, recondite truth to be discovered about whether these (zygotic clustering of) cells together constitute an organism or whether instead the organism begins to exist only later, when the cells lose their totipotency, become differentiated, and begin to be tightly allied both organizationally and functionally.[14]

One might have thought that the data given here remove the impossibility of 'discovering recondite truths' (or at least, some highly relevant facts deriving from scientific endeavours), and that my emphases, in the light of that biological information, highlight mere conjectural opinion. This mode of thinking is entirely redundant. Yet in pressing his point, McMahon invites his unsuspecting readers to rewind the biological clock by imagining themselves progressively reverting back to the 2-cell embryonic stage:

It is instructive to ask oneself when in this process of biological regression one would cease to exist. For my part, I find it impossible to believe that I would still be around when what we may neutrally designate as my organism had been reduced to a microscopic network of cells from which any possibility of consciousness had vanished.[15]

Why is it instructive to rewind the biological clock? Life is not a philosophical 'thought experiment' which sensibly allows us to regress, as McMahon suggests. We must reckon with the fact that evolutionary process including living matter, together with developing plants, animals and human beings, are all subject to the 2nd Law of Thermodynamics.[16] We progress in one, entropy-laden direction only, forever and irrevocably, onwards: that is time's arrow.

From the precise, empirical laboratory-based data outlined in previous sections, it is obvious that McMahon's use of 'totipotency', 'differentiated', 'tightly allied' are not only misplaced, but entirely inconsistent with the multi-gene processes that we now know are brought into play from the moment of sperm penetration. These processes give rise to the progressive, feed-forward effects which can only, in their orderly and regulated sequence, give rise to the blastocyst, and the primitive streak – and onwards to a full-term baby, and a mature adult.[17]

While McMahon consistently fails to define his use of the word 'organism', it seems from his regress (above) that he deems 'organisms' to be conscious individuals. In proceeding along this line of reasoning, we are logically brought to yet another contentious issue, whether the embryo-foetus or neonate are definable as 'persons' or 'organisms', and what is meant thereby. With that issue, we now confront an enormous literature, of which the most pertinent deals with ideas about 'persons', and the Utilitarian-Consequentialist calculus.

Who Qualifies as a 'Person' Thus to Enjoy Moral Respect?

To generalise, a considerable volume of philosophically-based writing asserts that the embryo-foetus, neonates and infants are not 'persons' and therefore have no rights whatsoever – whether a 'right to life' or any 'right' to be morally considerable: from that it follows that these 'things' so-defined can be destroyed. Previously (Chapter 7), I dismissed certain propositions, such as 'Was I ever a zygote?' or 'Is an embryo a person?' as questions irrelevant to an early embryo, or even a foetus.

Descartes first separated the immaterial mind from the substantial physical body. Later, John Locke defined person as 'a thinking intelligent being ... [with] reason and reflection ... in different times and places ... which it does only by that consciousness which is inseparable from thinking ... and essential to it'.[18] From that, we have notions of personhood that are entirely abstract, dependent only on outcomes of conscious activity, and thereby removed from any sound anthropological footings whether biological, sociological or theological.

Several 20th-Century writers such as Peter Singer and Michael Tooley frame mature personhood solely around desirable, arbitrary characteristics based on Locke's premises.[19]

Such tick-box approaches are based on subjective third-party attributions. But at what age are these abstractly-based items supposedly acquired and do they not change with the altering contingencies of life, of being an embryo, child, or mentally incapacitated, of growing old and demented? No single attribute is operative throughout any individual's life. More importantly, they inculcate dismissive attitudes towards immaturity, disability and progressive illness, facilitating beliefs that non-persons (by definition) can be killed without the least tingle of moral disgust or regret.

These views also fail to understand, critically, that basic conscious-arousal arises spontaneously during foetal life. On that capacity depends the later acquisition of those higher cognitive abilities which narrowly occupy those writers' thoughts. Others distinguish between 'life', and 'having a life'. The former implies either a foetal, mentally incapacitated, or near-cadaveric existence. The latter entails a worthwhile biographical existence integral to fully occupied lives of aspirations, preferences, memories, dreams etc., exemplary of a vivid, true personality.

One set of criteria (of many) in defining personhood (a-e below) was given by Mary Anne Warren, and we can usefully criticise it:

(a) Consciousness: for Warren this means the ability to perceive 'internal/external objects and events and *in particular* (my emphasis) the capacity to feel pain'.[20] But consciousness is not defined, nor even recognised as originating during foetal existence. Moreover, does loss of consciousness during sleeping, anaesthesia or from head injury extinguish personhood? Why are legal enquiries required into all post-surgical and post-anaesthetic deaths? For if 'personhood' has then been lost, such enquiries would be redundant – not to mention a death certificate! Incidentally, during these occasions normal adults are unresponsive to pain, although moral credibility is not thereby forfeited: neither with the extensive anaesthesia accompanying syringomyelia, chronic (lepromatous) leprosy, or tetraplegia.

Of greater importance is why Warren considers pain to figure so crucially as a criterion for personhood. Is she unaware, if not entirely oblivious, of the inherited disorder Congenital Insensitivity to Pain? These are normal people, exhibiting a full complement of other sensory attributes (vision, touch, hearing etc), and like all others within society, have full entitlement to belong to Warren's 'moral community'. The only difference is they should avoid anatomically dangerous positions or grasping very hot objects! And with a touch of better informed clinical guidance, they soon learn how to cope. So much for the moral relevance of pain. As a corollary, we should note that for an embryo-foetus to be considered of no consequence because it lacks 'pain sentience' would, likewise, seem to represent misplaced thinking. It is now mandatory that during foetal surgery or late abortions, sufficient analgesia must be administered.

(b) Reasoning: this is taken to incorporate the 'developed capacity to solve new and relatively complex problems'. We need to know what is meant by 'relatively complex'. Relative to what? And how complex? Surely there must be countless adults who lack that developed capacity, or have never been in circumstances where that capacity was required or even seen as essential for proper functioning. Yet are they not still regarded as proper persons, nonetheless?

(c) Self-Motivated Activity: such motivational activity is defined as 'relatively independent of either genetic, or direct external control'. Again, it is necessary to ask what *relatively independent* (again my emphasis) means here? And *relative* to what? And how can any type of activity – motor or cognitive – not be indissolubly tied to brain function and the necessary underlying motor or other cortical functions and controls? And how can a brain (and basic consciousness) come about without the *genetic necessity* for its construction, beginning *in utero* and continuing until the 25th year of adult life when its growth is ultimately finalised? If any motivated activity is independent of *external* control, then it can only be motivated *internally* and is therefore totally dependent on a functional brain and the neuro-muscular assemblies permitting these activities.

(d) Capacity to Communicate: Warren sets out her understanding of this capacity to be 'by whatever means...of an indefinite variety of types...not just an indefinite variety of possible concepts, but on indefinitely many possible topics'. I can only conclude that this stringent requirement would disqualify me! It sounds like the specification of a magnificent super-genius. I submit that even Einstein couldn't do all that – and certainly not your average departmental philosophy lecturer. Neither is this

capacity applicable to millions of ordinary people worldwide.

(e) Presence of Self Concepts/Self-Awareness: this seems to
apply to either 'individual, racial, or both' forms of awareness. In
commenting on this requirement for personhood, it is clear that
racial (or substitute tribal/religious/political) ideas, for example,
invariably become degraded into a self-conceptual awareness of
power, superiority, bestiality or destructiveness: witness the
many inter-tribal, ethnic cleansing African wars; the continuing
pseudo-religious divide in Northern Ireland; the usurping of
human rights and dignities (and especially the rape of women as
an instrument of war) in Africa and Arab countries and
elsewhere; and political dictatorship and domination of subject
peoples (Syria, Zimbabwe, Iran, North Korea).

Finally, arising out of Warren's claims, we note that in *not*
satisfying any of her criteria, the foetus – deemed a non-person –
has no right to life. She asserts that '…laws restricting the right
to abortion, or limiting the period during which an abortion may
be performed, are a wholly unjustified violation of a woman's
most basic moral and constitutional rights'. Is that so?

A similar approach to the actual burden of pregnancy is
offered by J.J. Thomson in her famous 'waking-up-attached-to-
the-world's-most-famous-violinist' thought-experiment.[21]

However, in real life and away from the smoke-filled, wood-
panelled rooms of philosophy departments, one simply does not
awaken to find oneself providing haemodialysis for a stranger in
renal failure (however famous – if *even* only for nine months -
and despite all the contingent ethical odds). Rosalind
Hursthouse effectively counters Thomson by asserting that
pregnancy does not confine women to bed – or attach the mother
to another individual never before seen – even if connected
through (placental) dialysis tubing.[22] Nevertheless, although
Thomson recognises that the foetus looks human (10-12 weeks),
she cannot assign any moral standing to it. Old habits seemingly
die hard!

We should, perhaps, wonder whether any female philosopher (including those like Mary Anne Warren or Judith Thomson) might think somewhat differently had they to undergo fertility treatment for a child so longingly, even desperately, desired. Would any other woman, pregnant after her third attempt at artificial impregnation, regard that precious object as a mere clump of cells, a non-person, or a parasite invading her body? Would she not rejoice at what was now going on inside her womb? Would she not even begin to dare to assign a name to that rapidly growing individual within her body, her very child to be? We can hardly perceive what unimaginable rapture would be engaging this woman's entire being when, on stepping out from the hospital into the street, the good news slowly began to sink in!

But we should recognise from these opinions that the embryo-foetus does not come into view solely as some kind of biologically – or socially – interfering parasite. One might have thought that in consenting to intercourse, most women would be intelligent enough to grasp the inevitability of pregnancy. Nor can the major paternal stake in that outcome be simply ignored: indeed, paternal genes produce the placenta! Moreover, as I explained above, the embryo-foetus operates under its own steam once instantiated through fertilisation, powered by the initial crucial interventions of the sperm, and from whence it will become a completely new, human individual. Its emerging bodily characteristics are entirely powered by the genome specific to that embryo-foetus.

Furthermore the embryo-foetus, although within the uterus, is an independent individual and not an 'intrinsic part' of the woman's body – contrary to much feminist propaganda. Therefore, contra Thomson, it is a controversial claim that women have 'rights' over their bodies giving them moral warrant to 'wriggle' out of unwanted pregnancies. In *not* being analogous to a maternal kidney or liver, the embryo-foetus exhibits individual separateness. This is exemplified with maternal-foetal blood group incompatibilities (Rh or ABO) when maternal mechanisms seek to destroy the developing embryo, analogous to rejection of tissue transplant mismatches. Conversely, abnormal embryonic (trophectodermal) moles, in reverting to chorionic carcinomas, invade the woman's body thereby threatening her very existence.

The 'Utilitarian-Consequentialist' Dimension

In proceeding, we must briefly identify basic features of utilitarianism–consequentialism (UC), since this principle also enters the ethical debate on abortion and infanticide. The major precept of UC is that the moral worth of any action is measured *solely by its outcome*. That is, fostering the greatest amount of happiness (JS Mill) or pleasure (J Bentham). This could involve trading off one individual against another, and since the ultimate aim is the achievement of the greatest good for the greatest number, actions are themselves of no consequence. We can murder, torture, abort and so on, provided the result achieves the required end. We can also tell lies, so long as the desired condition obtains: but that makes it difficult for anyone to rely on a utilitarian's honesty or trustworthiness.

Another conclusion about UC is that anything goes: it is neither one thing nor the other – 'it all depends'.[23] One problem is that 'greatest' pleasure and the 'greatest' number benefiting cannot be quantified and are idealistic rather than substantive ideas. Nor do we know, nor can we determine, at what time those great benefits will accrue. We should be aware of just how incompetent the human brain is in making sound, future predictions, especially when the evidence is slim or ill-defined.[24]

But to return to killing embryos, babies and infants. On the UC account, only the outcomes of actions are significant. But it is easy to see that if the previously aired exclusive definitions of personhood are aligned with UC, then the killing of embryos,

foetuses and infants has no significance because the result is good
– relieving women of unwanted pregnancies, removing a baby
with anencephaly because its parts are required for transplantation, or because a mentally-incapacitated child is occupying a
bed in place of another who could benefit. The outcome here for
UC is not particularly impressive.

As an addendum to the preceding, I draw attention to the
annual (UK) BBC television appeal (The National 'Pudsey Bear'
Appeal) for the continuing financial support of incapacitated and
disabled foetuses, babies and children (more than £32 million
raised by this public appeal in 2014). It is amazing how much
work, effort, expertise, long-hours freely given, and love pervade
the scene: such devotion exemplifies the convinced attitudes of
those involved in this ongoing massive country-wide project that
the disadvantaged *are* human, and *are* deserving of our utmost
care and support. The mealy-mouthed philosophical discussion
that abounds hardly seems to recognise – even understand – the
real world where these activities, *par excellence*, proclaim a
morally-intended set of values for those unable to articulate their
needs, provide for their own continued sustenance, navigate
unaided our complex world, or independently help themselves.

Final Thoughts on Abortion & Infanticide

In this chapter, an immense amount of ground has been covered
which might be considered to have been dealt with superficially.
That is inevitable, within a small essay, and particularly due to its
engagement with an extensive philosophically-based literature.
In attempting to avoid the thought-horizons pervading this field,
I have employed statistical and biomedical perspectives to illustrate the enormity of what goes on, and to evaluate the situation
from a legal, biological and theological position.

(i) Legal Issues: my argument, first, is that some knowledge of
the UK abortion Act (1967) is necessary. This enables readers to
grasp the scope of its legislation – what it allows, where, and by

whom. Of course, current hospital-based procedures protect
women. Yet, in retrospect, the legislation unsuspectingly opened
the floodgates, providing ready-made contraceptive services for
large numbers of young, unmarried women. One might well
conclude whether, before the Act became law, this exemplified
the huge repertoire of back-street abortion practice. More importantly, since the Act is not policed there is, on the fringes, illicit
medical behaviour. The staggering figures on abortion (two
million annually across the Western hemisphere) amply indicate
that it is the first port-of-call for terminations. Forty years on,
Lord Steele, who originally introduced this legislation into the
House of Lords (British Upper House), now laments its extent,
calling for greater responsibility towards contraception: an
'irresponsible mood' has emerged 'if things go wrong'.[25] That a
large cohort of young single women employs abortion-on-
demand demonstrates a practice not originally intended.

Without doubt, the genie is now out of the bottle and hardly
likely to return. As progressively higher proportions of abortions
are being carried out medically before ten weeks gestation, the
provisions of the Act are being effaced.

First, because at that early stage of gestation it is difficult to
identify precisely the maternal risks (mental or physical) which
the Act sought to control. The lack of critical review means there
is no compulsion by involved 'professionals' to exhibit moral
concern.

Second, we now live in an entirely different social milieu than
Britain in the 1960s when pregnancy outside marriage was
shameful. Indeed, current social attitudes neither encourage
greater respect for the embryo-foetus nor show concern about
abortion figures.

Third, because abortions are increasingly being effected by
drugs, there would seem to be no reason why two authorising
certificates are currently required to be furnished by medical
practitioners; many women assert that they are demeaned by

having to ask another (unrelated) person (doctor) for abortion.[26] The continued development of abortifacient pills may, indeed, render it unnecessary for women to attend their doctors: but that is for the future.

Fourth, government may still require knowledge about these affairs, although such practice would be more difficult to monitor since the illegal provision of drugs through the internet means that for many, abortion will be medically possible and hence continue at currently high levels, irrespective of attempted legal control. And since a medical termination does not produce a corpse needing burial, few people are likely to know, and certainly not governmental statisticians. The national statistics base does provide some insights into other reasons why a medically controlled death (Category E) of an embryo, foetus or baby may be required either resulting from congenital defects (~2,000 annually), or post-natal abnormalities, which seriously threaten continued, independent existence (>4,000 anomalies annually). Nevertheless, the combined total (~6,000) is markedly small in comparison with the total number of abortions carried out.

It is not easy from the published figures to determine whether those physically and mentally incapacitated children survive, or die naturally, either with or without assistance. Thankfully, society abhors child destruction, so the influence of UC thinkers has mercifully not become the norm. That point is made exquisitely clear in the widespread revulsion felt here in the UK against the deaths of a series of children, such as Victoria Climbié – from the culturally-mistaken (African) beliefs in spirit-possession and witchcraft, thus causing the family 'dishonour', to Baby P – who was persistently harmed and beaten to death in the most squalid of circumstances. But two issues, nevertheless, arise from these data.

First, that reproduction is a 'fragile' process: errors and defects occur throughout the entire process militating against those insisting on a perfect baby. All biological processes are shot

through with distortions, which may cause or require some form
of assisted death – whether *in utero* or in the neonatal period.

Second, the varied changes (social, therapeutic, and so on)
which have influenced abortion practice over the years should be
a warning to would-be legislators intent on changing the law
regarding assisted dying. In the next chapter, I draw on the
experience of our Abortion Act to indicate, in a more detailed
fashion, how future legislation concerning assisted dying could
be framed.

(ii) Biological Issues: in this and a previous publication, I
emphasised the biological issues that impact on the embryo-
foetus and beyond.[27] This is the second part of my argument. It
should now be clear, especially to those readers unfamiliar with
this territory, that the early embryo-foetus is not just a tangled
clump of meaningless cells. Sperm penetration of the ovum
achieves important, immediate effects. Once syngamy has
occurred, the new individual's genome becomes operative and
progressively induces cells to divide, thus to fulfil their
presumptive differentiated roles in the continuing, three-dimensional specification of the developing individual resulting from
more than 2,000 gene activations.

All biologists recognise that the origins and patterning of the
new individual are laid down, and specified for its future, at
fertilisation. Other later, arbitrary time-points employed by
bioethicists or philosophers fail to make sense: the failure of
those criteria should be recognised and peremptorily dismissed.
Often, it seems as if they are chosen merely to fulfil some pre-
determined stance that is illogical and flies directly in the face of
all the collective biological data.

Furthermore, against feminist perspectives, an embryo-foetus
is neither part of the woman's body on a par with her kidneys,
nor a parasite. For various biological reasons – developmental or
through incompatibility – either may invade, and even destroy,
the other.

There is a need to understand that other approaches (which I have touched on in this essay) help in characterising the developmental aspects of embryo-foetal life, and beyond, into adulthood. For example, in returning to earlier themes, I stressed that the embryo-foetus is not a new *life*, since the gonadal cells which fuse in constituting that individual are themselves living objects, reflective of the life-principle reaching back to the origins of life and of all living creatures on earth – in evolutionary continuity. That continuity extends even further back to the evolving, changing inorganic cosmos originating from the Big Bang onwards. We are all involved in continuity – in and through evolutionary and developmental changes.

Although there is no new 'life', there is a new individual – uniquely specified through its DNA profile. Yet DNA, in its Mendelian distribution throughout the species, is not the sole factor involved in foetal development. Environmental, or non-Mendelian 'epigenetic' influences are additional crucial factors regulating the expression of DNA (The Dutch Winter of Starvation and sociological studies from Sweden and Avon County, Bristol, UK) referring to the life-histories of grandparents in relation to their respective grandchildren (Chapter 7)).

The deep moral significance of these varied factors should be evident, and weighed carefully in the formation of new opinion, as well as the strategic planning of governmental initiatives for young children. These factors illustrate how three or even four generations directly influence the subsequent developmental profile of the growing embryo-foetus and its future post-natal life. Thus the embryo can be firmly welded into a physical anthropology, based 'vertically' through evolutionary time.

Finally, we are also subject to relationship and independence. Relationship is given and received in the intimate nexus of parent and family as community – what Martin Buber would have called *mutuality*; while independence is expressed by a unique genetically-moulded individual reacting to environmental cues while developing both structurally *in utero,* and psychologically his/her behavioural capacities and capabilities through time.[28] Such considerations bind each new uterine individual into a 'horizontal' social anthropology in intimate and usually loving contact with its parents and other family relatives.

(iii) Philosophical Issues: the third part of my argument is that we need to progress beyond the predominantly utilitarian-consequentialist debates that have dominated thinking in this area for several decades. This has engendered the freely liber-tarian approach to current abortion practice. There are, however, growing signs of unrest, in the UK and USA, on grounds that those freedoms have been obtained too cheaply; there being concerns (like slavery) that the pendulum has swung too far in favour of a come-as-you-will availability of the practice.[29]

Much of this libertine approach stems from woefully inadequate definitions of 'personhood' that are too abstract, lack adequately sound anthropological footholds, and rely too heavily on assumed cognitive abilities – self-consciousness, memories, future intentions, and interests.

For a moment, let us consider interests and desires, and the assertion by several philosophers that in lacking these, the embryo-foetus is not a proper person. But that argument is weak. It suggests that interests are solely based on subjects' own cognitive activity in developing their specific interests, and which are attractors of appropriate 'rights' and 'moral status'. But that is confused and fallacious. Parents confer an interest on their growing embryo by attending, for example, to the mother's health, her abstaining from alcohol, tobacco, drugs and avoiding environmental hazards and toxins: and later by ensuring the growing child has a suitable education. Alternatively, interests can be conferred on adults without anticipatory expectation, through the bestowing of society's awards, medals, or other honours. Moreover, an interest can be signalled for dead people (who, like the early embryo-foetus, lack interests and future intentions) through the posthumous award of the Victoria Cross, erection of a statue, or inscription on a commemorative plaque. Buddhists (and other spiritual beings) lack desires, as did Phineas Gage after an iron pole (in 1848) was blasted through his skull, yet they are still considered human beings.[30]

Professor Oliver O'Donovan lays down another criterion.[31] Starting from The Declaration of Helsinki through which those research programmes can be given proxy consent by parents or guardians, O'Donovan notes that such consent can be withdrawn. The withdrawing, on behalf of neonates, children or other disabled individuals, establishes the principle that any human subject, indeed, is morally considerable. That principle could, arguably, apply to an embryo-foetus at any stage of gestation *in utero*, either for experimental or therapeutic purposes. I noted earlier that when the foetus is being operated upon, the medical personnel concerned still carry full medico-legal responsibility for their actions, being liable for prosecution if catastrophe occurs.

Once we begin to ground either the embryo-foetus or infant into alternative anthropologies as indicated above, the landscape changes radically, and a moral concern for those particular human individuals now comes into view. This was the tenor of GJ Warnock's case.[32] He, in part, set out to develop an ethic compatible with the rightful claims of the inarticulate – whether *in utero*, or neonates and children, or others with brain damage or debilitating metabolic perturbations – thereby incapable of speaking up for themselves. Warnock was emphatic of their sufferability. He dismissed current perceptions of 'persons': instead, he reminded us that we, together, should act to recognise each other's vulnerability, and be morally responsive to its claims.

Readers should be aware of language usage, for example embryos referred abstractly as meaningless 'clump of cells', 'organisms' without interests or intentions, or unwelcome in an inclusively defined 'moral community' (whatever that means).

Hursthouse has similarly noted the limited vocabulary used in philosophical argument: acts are deemed 'right' or 'wrong'; or 'morally innocuous' or 'neutral' – that is, neither bad or good; or 'obligatory' or 'impermissible' – while agential activity is not directed to the person involved in its delivery, but deemed merely 'good' or 'wicked'.[33] She also indicts John Harris for employing 'embryo' in reference to all stages of gestation.[34] As she rightly points out, his text would convey an entirely different message if 'late-term foetus' or even 'baby' had been used instead of embryo throughout. Could this be seen as wilful deception and a concealed attempt at intellectual subterfuge?

Hursthouse raises another concrete example that if aliens had browsed many of these texts, they would gain no insights whatsoever about the specific, behavioural characteristics of human males and females, the nature of sexual intercourse, the duration of pregnancy, the inherent dangers of childbirth, that abortion or infanticide mean the very killing of an innocent human being – whether lively or incapable of survival, neither the meaning of 'family' nor of relationship – even of love.[35]

Despite all that has been written, we should recognise that the correct solution to abortion or infanticide has not been achieved, and that there are other approaches which offer more humane attitudes towards some of the most unprotected members of the human species.

(iv) Theological Perspectives: from those derived anthropologies, a rich tapestry from which the undoubted moral significance and value of the embryo-foetus and infant can be established. Such deliberations, as set out above, cannot themselves offer 'proof' for the necessary theological status or spiritual value of these early human beings, any more than appeals to Design, or the Anthropic Principle, can inevitably lead to belief in the Godhead – but they are strong pointers in that direction.

Nevertheless, the debate over abortion and infanticide has to be fought in an intensely secular forum: arguments based on 'made-in-God's-image' or 'sanctity of the body' cut no ice in that milieu. Yet my having grounded the developmental trajectories of nascent human beings within physically and socially based anthropologies, and having established some of the moral implications arising, one might expect the opinions of more thoughtful secularists to change.

So to the final (fourth) part of my argument. How can a theologically-based anthropology be determined?

An individual, as proposed by Professor John Zizioulas, comprises the 'One who is…', thus aligning with Professor Oliver O'Donovan as a 'someone who …', each referring to the essence ('is-ness', or 'is-ing', or 'being') of that individual, both in historic continuity, but also locally related with others in loving intimate familial and communal settings.[36,37]

At first sight, this is a subtle definition. Yet it incorporates, at any time in that historical, developmental and relational existence, both the physical and metaphysical characteristics unique to every individual (Chapter 6).

There is no doubt that biologically, genes do mark out individuals as distinctive and distinct from many parental characteristics. But neither genes, nor the origins of brain functioning *in utero*, acquisition of interests, desires or self-consciousness, nor the later attributes of psychological and cognitive abilities, properly encapsulate the entire, innate essence, or 'is-ness' of every being. The 'one who is' defines with subtlety the precise specification of each individual, physically and metaphysically, at any time during the being's existence, from conception to the grave.

Proceeding from theological perspectives, O'Donovan considers that although 'person' is a non-biblical term, it assumes Patristic significance in respect of Jesus whom the Fathers wished to emphasise was, on Chalcedonian principles, one substantial person, of two natures.[38] Thus any person or individual is

anterior to nature, the particular properties exhibited by that
substantial person (Latin = *persona*). The Greek word 'hupostasis'
also implies the reality of substance underlying (or 'standing
below') those properties characterising each subject (or
individual, used as a noun, and not in its adjectival sense).[39] In
both biblical testaments, coming into being is a vocation given by
God the Father, and establishing importantly a 'someone who…',
that is, an individual (rational substance) with a history, a name,
and, within that history, given and coming into a relationship
with one's forebears and future progeny, and hence
demonstrating a life-history of continuity and of change.[40] There
is a familiar ring here relating to anthropologies given above – of
'vertical' continuity and of 'horizontal' relationship.

Here, as O'Donovan notes, these are new beginnings fulfilling
hopes prophetically foreshadowed but, especially in the birth of
the Word, endowing all births with holiness and sacredness in
reflecting God's purposes in and for his created order.[41] The
begetting of new, genetically-unique individuals is suffused by
the power of the Spirit, in evolutionary continuity with the
origins of living matter, and that from the progressive
evolutionary conformation of inorganic matter from that
moment of Creation.

We are, in analogy with Jesus' relationship to the Father,
begotten of ourselves, not exterior 'things' or man-made
artefacts. In not making, we employ the living vehicles (egg and
sperm) available to us, and naturally provided in (evolutionary)
continuity for the purposes of begetting. Such begetting by
humans is subject to the laws of natural existence, as against
those that merely obtain in the creation of artefacts, and the
manipulability of the materials employed in their creation.

The rule here is not the abolition of limits through sheer
technological ingenuity and inquisitiveness, a rule which
apparently permits women to declare themselves inviolable to
pregnancy and what it entails – biologically and morally – if not

theologically. On the contrary, the rule that governs begetting is love, manifested and achieved through O'Donovan's percept of neighbourliness, or GJ Warnock's sufferability.[42,43] Upon those criteria is based the absolute moral considerability that Warnock sought for the embryo-foetus, neonate, infant, and the brain-damaged child – and from which the warrant, which notably confers moral considerability upon those unable to speak up for themselves, arises.

Whither Abortion & Infanticide?

If it were felt that abortion levels should be substantially reduced, we would need a far more explicit definition and acceptance of the moral status of the embryo-foetus, and its value as a human subject, than presently exists. And that tighter view would have to pervade society as a whole, not simply at local or national level. That would demand a dramatic re-ordering of attitudes: indeed, current figures for abortions across Europe indicate just how substantially difficult this undertaking would be.

For example, if we were to concede (reasonably) that about 5-6,000 abortions (annually in the UK) are necessary because of congenital defects, other variously acquired intra-uterine problems, and for certain other reasons, we would still be left with ~190,000 abortions annually performed on demand purely for contraception. Warren does, rightly, ascribe responsibility for unwanted pregnancies to those acting without restraint.[44] That responsibility would be all the more focussed if, generally, society regarded the embryo-foetus as morally significant, if not of spiritual value.

From other disease perspectives, the duty to take medications is vital, either for disease control (eg. epilepsy, diabetes, transplant rejection) or attempted eradication (infection or malignancy). Therefore, the avoidance of an unwanted pregnancy requires the conscientious taking of tablets or use of other preventatives thus ensuring, given the possibility of

intended sexual activities, that full protection is guaranteed.
There is no doubt that the reality of pregnancy concentrates the
mind – but *after* the event. It would be far preferable if the
expectancy of pregnancy would concentrate the mind *before* the
event.

Next, it follows that we should engage with those calling for
improved contraceptive advice – particularly towards
schoolchildren – and its effective practical deployment.
Currently, it is unclear whether so-called contraceptive advice in
schools is working effectively. Those calls emanate, in part, from
those who see as unacceptable the increasing rates for very early
teenage (12-16 range) pregnancies and abortion as reflections of
the ineffectiveness of contraceptive teaching. At present, that
surely is a cry from the wilderness. For these young teenagers, to
have engaged in intercourse could well be seen as a 'badge of
office'.

How can we impose better informed standards of
contraceptive advice and practice, when for others, the embryo-
foetus is dismissed as a nonentity; a position legalised by the
Warnock Report and its conclusion that persons only come into
view after 14 days, and by philosophers who regard killing an
embryo-foetus as of no consequence whatsoever because it is, by
their definitions, a non-person? Or by those who think women
are demeaned by having to ask a doctor for an abortion? Or by
those who assert that it is part of a woman's autonomous,
absolute right to 'wriggle' out of pregnancy and dispose of any
embryo-foetus, when and where? Or by society that cares little?

On another front, the national database allows us to respond
in a measured way to those who declare all abortions to be
immoral, or even amounting to murder.[45] The use of such an
emotive word does little to bring the issue into the cooler light of
reason: nevertheless we cannot escape from any abortion other
than constituting the taking of the life of a human being.
Reference to the published figures indicates that ~1% of post-

natal deaths are inevitable due to natural defects simply incompatible with life, due to the effects of major surgery, infection, immune depletion, or that no (current) sensible medical intervention is possible, and so on. And for many parents, these losses would be irreparable. Indeed, in all such circumstances, we should apply the 'Biggar Test': that is, in realising that many of those embryo-foetuses and infants are doomed to die, whether *in utero* or during neonatal life and beyond, we should be respectful of, or exhibit 'sorrow' towards those inevitably lost individuals – before and after birth.[46] But despite current outcomes, the Law on abortion, as it stands, is here and within our midst. If we were ever to attempt turning the tide on the present number of abortions, we would need a new (moral) crusade. First, by convincing everyone that the embryo-foetus is a real part of humanity, tied by close family bonds into the larger human family, and thus to be morally respected and even to be regarded as a holy thing given by God, and therefore not to be extinguished with impunity. Second, to pursue effective campaigns ensuring that contraceptive practice is more intensively taught, understood and conscientiously applied. Neither of these challenges is likely to be met soon, nor easily accomplished.

Indeed, we might all fear for the future.

* * *

1. Fradd CR, 1985. An Introduction to the History and Present State of the Law Relating to Abortion in England. In: Channer JH (ed), *Abortion & the Sanctity of Human Life.* Exeter: The Paternoster Press 140.
2. An error in judging something's worth: here, presumably, the worth of any pregnancy.
3. A further amendment, S.37 (Human Fertilisation & Embryology Act, 1990), reduced the 'therapeutic' time limit

to 24 weeks gestation (on Categories C, D) while removing
the time limits on the remaining Categories A, B, and E.

4. 1939 1KB, 687: Rv Bourne.

5. The sources of data in this chapter are derived from the
   Office of National Statistics (ONS), London; The Johnson
   Archive 2010; Statisches Bundesamt, DDR 2010; Wikipedia.

6. Human Development Anatomy Center (National Museum
   of Health & Medicine, Washington, DC).
   http://www.lifeissues.net/writers/irv/irv_123carnegiestages
   1.html;
   http://nmhm.washingtondc.museum/collections/hdac/stage
   1.pdf]

7. Whitaker M, 2005. Syngamy and cell cycle control. In: Myers
   RA (ed), *Encyclopaedia of Molecular Cell Biology and Molecular
   Medicine*. Weinham: Wiley Verlag, 1-35.

8. Tran L, Hino H, Quach H, et al, 2012. Dynamic microtubules
   at the vegetal cortex predict the embryonal axis in zebrafish.
   Development 139: 3644-3652.

9. Ford N, 1988. *When did I begin?* Cambridge: Cambridge
   University Press.

10. Singleton J, 1994. Ethical principles at the beginning of life. *J
    Reprod Infant Psychol*, 12: 139 (on two occasions); Greenfield
    S, 1998. *The Brain – A Guided Tour*. London: Phoenix 120;
    Thomson JJ, 1971. A defense of abortion. *Philosophy & Public
    Affairs* 1: 47-66 (48); McMahon J, 2002. *The Ethics of Killing*.
    Oxford: Oxford University Press 25.

11. Green R, 2002. Part III – Determining Moral Status. *Am J
    Bioethics* 2: 20-30 (22).

12. Lorthongpanich, Doris T, Limviphuvadh V, et al, 2012.
    Developmental fate and lineage commitment of singled
    mouse blastomeres. *Development* 139: 3722-3731.

13. Vassena R, Boue S, Gonzalez-Roca E, et al, 2011. Waves of
    early transcriptional activation and pluripotency program
    initiation during human preimplantation, *Development* 138:

3699-3709.

14. McMahon J, 2002. *The Ethics of Killing.* Oxford: Oxford University Press 28.

15. McMahon, *Ethics* 29.

16. Prigogone I, 1996. *The End of Certainty: Time, Chaos, and the New Laws of Nature.* New York: The Free Press 17.

17. Marsh MN, 2012. *The Moral Status of the Embryo-Foetus: Biomedical Perspectives* (Ethics #166). Cambridge: Grove Publications.

18. Locke J, 1995. *An Essay Concerning Human Understanding.* New York: Prometheus Books 246-250.

19. Gordijn B, 1999. The troublesome concept of the person. *Theoret Med Bioethics* 20: 347-359.

20. Warren MA, 1973. On the moral and legal status of abortion. *The Monist* 57: 43-61.

21. Thomson JJ, *Defense* 49.

22. Hursthouse R, 1987. *Beginning Lives.* Oxford: Blackwell 203.

23. Hursthouse, *Lives* 134-5.

24. Tversky A and Kahneman D, 1974. Judgement under Uncertainty: Heuristics and Biases. *Science* 185: 1124-1131.

25. Ward L and Butt R, 2007. *The Guardian,* 24 October.

26. Hewson B, 2002. SPUC and the morning-after pill saga. *New Law J* 152: 1004.

27. Marsh MN, Moral Status 13-21.

28. Buber M, 1959. *I and Thou.* Edinburgh: Clark, 2nd Ed.

29. Kliff S, 2011. *The Year of the Abortion Restrictions.* The Washington Post (29 December) notes that 83 laws restricting access to abortion were passed in US, and that anti-abortionist State Governors have increased from 10-15.

30. Neylan TC, 1999. Frontal Lobe Function: Mr. Phineas Gage's Famous Injury. *J Neuropsych Clin Neuroscience.* 11: 280-283.

31. O'Donovan O, 1984. *Begotten or Made?* Oxford: Clarendon Press 56.

32. Warnock GJ, *Morality,*

33. Hursthouse, *Lives* 219.

34. Hursthouse, *Lives* 129-130.

35. Hursthouse, *Lives* 180-181.

36. Zizioulas J, 1993. *Being as Communion.* Crestwood: S. Vladimir's Seminary Press 27-65; idem, 1999. On becoming a Person. In: Schwöbel C & Gunton C (eds), *Persons Divine & Human.* Edinburgh: Clark 33-46.

37. O'Donovan, *Begotten* 59.

38. O'Donovan, *Begotten* 49-50.

39. O'Donovan, *Begotten* 50-51.

40. O'Donovan, *Begotten* 59.

41. O'Donovan, *Begotten* 55.

42. O'Donovan, *Begotten.* 5-6: 13.

43. Warnock GJ, 1971. *The Object of Morality.* London: Methuen 28: 87-93: 152-156.

44. Warren, Moral and legal status 50.

45. Foster J, 1985. Personhood and the Ethics of Abortion. In: Channer JH (ed), *Abortion & the Sanctity of Human Life.* Exeter: The Paternoster Press 31-53.

46. Biggar N, 2004. *Aiming to Kill: The Ethics of Suicide and Euthanasia.* Cleveland: The Pilgrim Press 114.

Chapter 10

Assisted Death & Assisted Suicide

At the time of writing, there is no legalised assisting of suicide in the UK. There have, however, been repeated calls for such legislation which has been promoted through our upper parliamentary chamber, The House of Lords. Furthermore, unlike other legislatures which have passed appropriate laws, the approaches in this country through Lords Joffe and Falconer, have been based on an anomaly of the Suicide Act. The Act was modified in 1961 such that while the act of suicide itself was decriminalised, the assisting of a person to commit suicide remained operative.

Now, in light of a perceived need for assisted suicide and assisted dying, that remaining part of the Suicide Act is seen as the stumbling block. It is that block, with additional provisions, which the efforts by their Lordships have sought to change.

These repeated attempts, which so far have been voted down, are associated with names of key subjects who (probably being supported by pro-death organisations) make good newspaper headlines with either their current sufferings, or anticipated sufferings. Thus Tony Nicklinson who died recently from locked-in syndrome, Debbie Purdy with MS, and others whose deaths in Switzerland have been well publicised, continue to keep assisted suicide (AS), together with Falconer's recent report, firmly in the public eye.[1,2]

For Christians, certain notions such as the killing of human beings (Sixth Commandment), perceptions of 'sanctity-of-the-body' – or 'sanctity-of-life', 'man-made-in-God's-image', 'life is a gift from God', or 'God gives and God takes' seem gravely inhibitory towards any kind of killing. Despite that, Christian opinion is divided on the issue.[3] Moreover, the ethical literature on

AS is both extremely extensive and daunting, and fails to provide uniform answers. This does present a dilemma, making it difficult for the average person to understand what is really going on, to mount a determined response, and thus to act with effective challenges. And yet Christians should be able to articulate coherent responses, and to get a grip on what is at stake here.[(4)]

Reviewing Lord Falconer's Proposals on Assisted Suicide

Following attempts to change the law by his predecessor, Lord Joffe, Lord Falconer introduced a clause in the Suicide Act to absolve accompanying persons (travelling to Dignitas in Switzerland) from prosecution.[5] This initiative was defeated. Notably, we should observe that suicide assistance (over 200 cases to date) has never attracted retrospective prosecution by the UK authorities [The Director of Public Prosecutions].

Undaunted, Falconer then set up his ad-hoc committee of enquiry (Falconer-Demos Report) into Assisted Suicide.[6] This again rehearsed his mantra that the law is 'inadequate and incoherent', adrift of 'public opinion' which supports assisted killing, and, in having 'been overtaken', is in need of 'updating' thereby allowing those wishing to end their lives to be helped by others without fear of prosecution. Such individuals must declare soundness of mind and be certified as 'terminally ill' by 'two wholly independent doctors'.

His current Bill 24 (HoL, 2013) is progressing through the Upper House.[7] It is, *notably*, a much watered-down version of his earlier, but determinedly rhetorical stances. The Bill now promotes the more simplified notion that two criteria must be fulfilled and certified: first, that of a mentally and physically competent individual who, with a firm and settled intent, could administer the lethal treatment without assistance, and second, a person deemed to be 'terminally ill with less than six months to live'.[8] Such persons should be allowed to take their own lives given the necessary prescribed lethal cocktail.

We should notice that these individuals, now, must be *within six months* of death, rather than *one year*, as had always been proposed on earlier occasions. This change might be intended (merely perhaps) to suggest an easier outcome in forecasting the probability of an individual's imminent death. Second, with specific reference to the Abortion Act, two doctors are required to make the declaration, 'independently'. Much thereby hangs on those 'independent declarations'.

In the Demos Report, Lord Falconer assures us that any prospective legal framework would be so tight and with "robust upfront safeguards" that it would not be easily thwarted. It is difficult to comprehend how a previous Lord Chancellor of England should think it possible to devise a law incapable of being by-passed by those intent on doing so.[9,10] Indeed, we might even review Lord Falconer's record of repeated attempts to achieve his ends through changing, or circumventing the law, by his various relentless strategies outlined above.

My argument asserts that Christians should beware the Falconer-Demos Report (and other cognate proposals arising) as vehicles embodying cogent reasons for law change. The basis of my argument derives from critical analyses of the Abortion Act (1967), which can be observed from its progressive *manipulation* over the years, as well as further possible *attenuation* by future and unpredictable forces, which are never forseeable when law is formulated. The natural history of the Abortion Act is thus employable as a source of great ancillary wisdom, whether or not alteration of legislation for AS ever comes about.

I have been accused of equating abortion with end-of-life issues. That is completely false. As is evident from what I have written, I regard a pregnancy as a sacred event, while there are theological grounds for allowing the disposal of an end-stage body, provided the criteria are correct. Since the Abortion Act has been in play for almost 50 years, we can surely elucidate some key observations from its use which could inform the formation of an AS law, thereby avoiding the errors inherent in the former.

From that, I am offering new initiatives albeit founded largely from a bio-medical perspective – in other words, observing in part what doctors cannot – either through lack of training, of experience, or through the sheer impossibility of what is being asked of them. This is a different viewpoint – unlikely to be foremost in the thoughts of those either proposing, or worried by, far-reaching changes to the law. I believe that other sterner safeguards are available to ensure that AS and even assisted deaths (without any further legal reform of the UK Suicide Act) could be pursued ethically, and thus as honestly as it is possible to ensure. These approaches should help to assuage either theological or practical difficulties for Christian people caught up in these terrible dilemmas.

Using Abortion Data to Anticipate & Avoid Future Deviations

As I have stated previously, we can derive wisdom from the ways in which the Abortion Act (1967), has been deployed. My case, based on the emergent shortcomings of the (Abortion) Act, both in context and in its practice, concerns whether legal changes relating to Assisted Suicide are, indeed, the primary requirement here. And, if that is *not* the case, what other strategies could be envisioned in order to guarantee fair passage to those assisting the victim. In fact, it is a pity that after Lord Joffe's attempts, a completely different approach was not attempted, whose scope could have been imaginatively widened to include others, in addition to those able (by current require-ments) to end their lives.

(a) The Ethos of the 1967 Abortion Act: briefly, Lord David Steele's Bill required that abortions should be undertaken in registered (UK) hospitals by competent surgeons in clean condi-tions. The obverse was to outlaw lethal, back-street abortionists. But despite achieving those aims, within three years another

completely unforeseen outcome, whereby the Act would now be understood and acted upon, arose.

That outcome, *with all restraints now abolished*, resulted in a rapidly manifest 'slippery slope' maximising within three years up to 200,000 abortions per annum. And a rate consistently maintained: thus, during the working of the Act, approximately eight million (UK) embryo-foetuses have been aborted. Comparatively, resulting from similar, albeit non-identical legislative enactments, two million abortions across Europe, UK and North America are performed annually.

(b) Categorising the Provisions of the Abortion Act (1967): five operational categories were stipulated by the 1967 Act.[11] According to ONS, apart from Category E (malformations) and Category D (relating to the welfare of pre-existing children (each having remained constant at 1000 abortions per annum), all certifications since 2000 have virtually ignored A and B, and employed category C, that pregnancy continuing beyond 24 weeks would be more harmful to the physical or mental well-being of the woman.[12] Clearly, the discernments specified, in accordance with the Act's provisions, of evaluating each patient thoroughly, have progressively been eroded by doctors. Moreover, the use of category C can never be wrong.[13] Additionally, we know that ~50% of all UK abortions (60% in Scotland) are now effected with medical abortifacient drugs rather than surgery, and, importantly, *before the ninth week of pregnancy*. Clearly, abortion is now the preferred 'contraceptive', despite other effective preventions; is being completed at a *very early stage*; and with the convenience of tablets.

(c) Being Wise after the Event: from these developments, what discernments can we draw which, prospectively, might be useful in formulating legal permissions for AS?

First, that the original stipulations of the Act have come to be viewed in a different light, subject to new social attitudes, newer therapeutic options, and altered operational approaches by doctors.

Second, greater worries regarding the widespread use of category C relates to how two independent doctors – by law, and supposedly 'in good faith' – could possibly determine, at this extremely early stage of pregnancy (before nine weeks), and involving large numbers of fit, young girls (predominantly unmarried), that the continuation of pregnancy would (even might) endanger their physical or mental well-being.

Third, are we to suppose that there has never been disagreement, even on one single case, among the two supposedly independent certifying doctors regarding any of these eight million (previously) aborted foetuses? And what quality of examination by each doctor has been taking place over these years?

Fourth, there is more disturbing news. Several recent 'spot' checks of some abortion clinics revealed supplies of pre-signed forms, a malpractice immediately referred to the General Medical Council's Disciplinary Committee. Obviously, *only one* (the second) *doctor* could have supposedly been acting legally (maybe, perhaps) in these circumstances.

Fifth, ONS figures reveal that in 2009, the gestational details and grounds for abortion were not included on eighty-four forms but tendered *after the event,* while in 2011, authenticating forms lacked the grounds for abortion in twenty-two, and gestational details in forty-two, cases. All those abortions were technically illegal. Clearly there are many doctors who simply cannot be bothered to sign forms at all, or to provide accurate details: while surgeons allegedly terminate despite incomplete referral documentation.

Sixth, as abortion is now legal it is demanded outright. Doctors comply without too much regard either for the patient or the law.

Seventh, since the provisions are legal, there can be no whistle-blowing.

Looking Forwards: Adjusting Legal AS in Light of Abortion Practice

From these accumulated data, it is not too difficult to see how some parts of the medical profession have somewhat bent the rules over the years, showing considerable disregard for the Abortion Act's requirements. One overriding difficulty, in my view, comes from lack of proper supervision. The police are not involved, so there are no proper constraints on rogue behaviour.[14] Performance has become an exercise in rubber-stamping, with little thought involved, engendering a progressively careless concern for, and attitude towards, the law.

Nowadays it is most unlikely that anyone is conversant with the original aims of the Abortion Act. Indeed, it is not uncommon for the provisions of any Act to be completely forgotten with time, thus to be eroded and over-ridden. As time progresses (almost fifty years), it is clear that society changes, while at the same time, any legal provision – like a photograph – becomes historically ossified from origin. Thus, once a law is enacted, modification is extremely difficult, so that erasure from the Statute Book is virtually impossible.

These changed circumstances, entirely unforeseen in 1967, should caution our approaches to the possibility of legalised Assisted Suicide – now, and especially for the future.

(a) Examining the Role of the Certifying Doctors Involved & Other Issues: Most manipulations (and this needs to be realised with respect to Assisted Suicide) rest entirely and heavily on the medical profession. Can, therefore, the profession (through Lord Falconer's proposal for two, independent registered practitioners – as per the Abortion Act) be entrusted to perform the sacred job of certifying those who request AS? And can we be reassured that no malpractice would be employed in securing those ends, irrespective of the so-called ideals of professional integrity and the dispassionate approach?[15] From what we have seen regarding abortion, the answer might well be 'no, it cannot'.

Neither is it quite clear what Lord Falconer means by two 'wholly independent' doctors, and why he thinks that is an important or even useful necessity. It would certainly be inimical to a patient's assessment for AS if, as Baroness Finlay, a professor of palliative care, succinctly put it, the two doctors involved had just been medical students on the day previous.[16] Clearly, they should be very senior doctors (of at least 10-12 years standing), preferably a psychiatrist with experience in end-of-life matters, and a palliative care physician. These would evaluate the case, even though brought forward (in part) by the local doctor.

But second, despite working locally, what logical reasoning necessitates – or even demands – that the two doctors must report independently? Lord Falconer, obviously influenced by abortion practice, appears to think that there would never be disagreement. That is hardly likely, as we explored above. Alternatively, most sensible people might concur that any disagreement occurring over time from assessments done properly with the interests, not to mention the life, of the patient under consideration should, indeed, be publicly aired – and rightly so – to and for the advantage of that subject being reviewed.

Recent surveys by professional medical organisations (Royal College of Physicians, Royal College of General Practitioners, Palliative Care Association, etc.) showed high percentages of doctors unwilling to be caught up with the death of patients. It is therefore difficult to see how the 'two doctors' would be recruited from among the ranks, or even from one location relevant to the patient. One would hope to avoid 'doctor-chasing' as is alleged to occur in Oregon State, for example.

Next, 'terminally ill' and 'death within 6 months', demand much further critical scrutiny. These are not only meaningless and hopelessly vague within the context of a presumptive, far-reaching alteration to the law, but are also impossible to define. While we all have an inkling of what terminal implies, that does not logically guarantee any doctor's ability to implement such an opinion. That is an impossible task, and no medical school curriculum teaches the criteria underlying such imagined capabilities – whether within a six or twelve month future period.[17] Those debating these issues in The House of Lords felt most uncomfortable with these provisions, their indeterminate definition and hence how they could ever be implemented in practice. One might wonder if these artificial time restraints played any helpful role in the deliberations.

(b) *The Problems of Assessing Mental Status:* Another very vexed question concerns the proper assessment of the mental state of these patients, as was repeatedly raised in the Lord's debates. This is by no means easy, since the subjects' moods and feelings are notoriously changeable. One of the most difficult diagnostic decisions is determining the presence, and depth, of clinical depression. Such assessments may not be possible on a single visit, and since the mood level in depression is daily variable, a very informed decision would be needed before any forms were signed – and anyone's life ended. Therefore, it is evident that time for deliberation is required in these very difficult end-of-life decisions, and that they should be carried out by experienced hands: and that must be acknowledged. With the currently proposed Bill (2013), there is no discernible differentiation between depression, and the separate condition of 'demoralisation', now progressively defined over at least twenty years.[18] Third, a sense of *hopelessness* appears to engender suicidal ideation in patients with end-stage cancer: that also needs recognition and dealing with.[19,20] Without going into detail, there are also the problems of anxiety, delirium and cancer-related encephalopathy.

According to Baroness Hollins (a former President of the Royal College of Psychiatrists), only 6% of Oregon psychiatrists claim ability in assessing the mental states, competence to make judgements, and needs of end-of-life subjects. Furthermore, medical competence is not uniform, and that variability in clinical competence as well as experience in these matters must be borne in mind. This would suggest that end-of-life psychiatric practice should become a major sub-speciality within this clinical discipline, especially as the number of deaths of this kind may continue to grow.

In continuing, she voiced much concern that doctors generally are very poor at recognising depression, and greatly underplay the possibility of depression, especially in end-of-life circumstances. Depression, in these situations (such as cancer, motor neurone disease, HIV-AIDS), is strongly associated with a desire for hastening death. And since these feelings are volatile, several visits to the patient would be mandatory in order to exclude these overriding possibilities.[21] Depression is commoner in those with physical disability, especially when associated with chronic pain, reduced mobility, or poor social support. Again, depressive symptoms can overlap those of physical illness, such as fatigue, lack of interest in previous pursuits, poor sleeping, and emotional liability. Manifestations of mental incapacity differ in those of higher intellectual status from those at the lower end of the spectrum, or with autism in which there is social withdrawal, poor communication, sleep alterations – all of which can simulate depressive episodes.[22,23]

Despite all these rather massive clinical problems, we also know that when mental problems are expertly dealt with, and resolved, the majority of patients no longer wish to take their lives. Much, therefore, hangs on the necessity of evaluating this aspect of patient management very carefully. Once someone has died, there can be no going back.

(c) The Required Legislative Safeguards & Oversight of AS:
The loopholes in the Abortion Act clearly demand strict oversight, and very sensibly, the Lords strongly demanded ultimate judicial oversight, through the Family Division of the High Court of Justice. This court is served by about 19 judges, and

works in regional centres. Therefore, there would be ultimately effective legal supervision acting, to a great degree, as a deterrent to malpractice. Furthermore, since the courts take the final decision, that removes all responsibility from doctors for actual assistance with death. This would go a long way in assuaging public opinion about the role of doctors in these matters.

My own view is that the initial momentum in deciding on a case of this kind is for appropriate parties (the patient, the local doctor, a solicitor, or a family member – especially if there are differences between the patient, family and the doctor which do arise) to be freely authorised to approach a local, legally empowered panel. This would be chaired by a senior legal person, and they would be permitted to use other relevant people, including an end-of-life specialist, solicitor (lawyer), nurse, pastor, or social worker. Here the case would be put and sympathetically heard, and then dealt with in a relaxed manner, away from the initial formality of court proceedings.

This panel would have powers to co-opt other cognate expertise as crucial adjuncts to evaluating the application and the competence of the patient's request. It is quite evident that the input of relevant specialists (pain relief, cancer management, etc.) has a vital role to play in assessing treatment to date. Given significant changes to management, there is often a change of heart, as repeatedly shown in other jurisdictions where AS is legally permitted. Hence the reason why all medical assessments, and views and opinion deriving from them, should be publicly open for general discussion among those involved in the death under consideration.

For Falconer, the 'patient' seems to exist *in vacuo*, rather than being located in geo-physical space: and that's important. Is the subject at home? Or residing in a residential utility? Or a care home? Or a hospital (geriatric) ward? Or in a hospice which practises effective Palliative Care? These determine the five possible origins from which a request for AS could arise. On those grounds, it could not possibly always be the initial responsibility of the GP, as universal prime mover, as Falconer seems to assume.

However, to prevent slippage, and the kind of malpractices that we have seen over abortion, requests for AS should be referred *before* anyone's life has been taken. The combination of retrospective analysis and gross irregular post-event form-filling, as has demonstrably occurred over abortion, would be totally unacceptable with AS, (i) in principle; (ii) for ensuring public confidence; and (iii) and most importantly, in offering reassurance to the subject. There is a need for the ultimate sanction of legal proceedings to remain in force against any malpractice: the Suicide Act would also remain intact.

Furthermore, such sanctioning would do much to calm those who thought their lives were at risk when unable to resist such pressures, whether perpetrated by doctors, family or social workers. Others, including senior social workers or religious ministers, should be involved in the assessment, especially if they have expertise in hospice care and allied decision-making. Those with experience of ethics committees know that good decisions are always possible, especially when it involves individual lives. Once the decision is upheld, following the conduct of these very important clinical assessments, the High Court could then be approached, the reasons for this considered approach given, and the necessary paperwork signed off. At this stage, the police, the Chief Coroner for England, and the Office for National Statistics would be informed.

Finally, all undertakers should be on notice that it would be illegal to accept any body for burial unless first accompanied by the appropriate certification by the High Court judge involved.

At the Ending of the Day
In finishing, I refer back to Lord Falconer's mantra that the Suicide Act, 1961, is incoherent, unworkable, and out of touch. I

resist that, on the grounds that his claims are (1) *incoherent* – because laws remain fixed while society's needs change over time; (2) *unworkable* – because his definitions and the details of procedure are vague, only superficially thought-out, and uncertain, while the dependent role of the two certifying doctors is totally unclear or their actions imprecisely defined; and (3) *out of touch* – because there is a need for deterrence in order to overcome, and punish, wrongdoing. This must be the case in order to restore and maintain public opinion, and especially because – at the moment – corrupt behavioural 'norms' are currently endemic throughout society, at all levels, from government (even the Lords, too!) downwards.

Buttressed by my illustrations of how the Abortion Act has been (mis)deployed, I believe we can devise procedures that, whether Assisted Suicide becomes law or not, embrace sympathetic examination of certain individualised cases that could be handled locally but under the ultimate jurisdiction of the Court, and with the necessary overriding deterrence of police oversight, and prosecution.

Conversely, it should be realised that a blanket re-arrangement of the Suicide Act itself, would probably have the effect of alienating many from diverse religious backgrounds and persuasions. My own view is that a completely new law should be devised, so as to cover all eventualities, without the need for further additional legal additions to be debated and enacted over time.

The alternative strategy, adumbrated here, embodies a permissive, compassionate approach fashioned to quell anxieties held by those with firmly held Christian or other religious viewpoints on assisted suicide or assisted death, given that the strategy was seen to be working. We note that Christians (and those of other religious persuasions) have to live with various legally sanctioned acts, which interfere with their beliefs or way of life. At least they could be reasonably satisfied that a law-based approach to AS was in force and which, more importantly, demanded strict adherence and oversight. During the subject's last moments within this now more relaxed approach, there could be a conjoined celebration, even within a eucharistic setting as Professor Paul Badham envisions.[24] Indeed, his ideas provide the foundations for some very much-needed and renewed Christian understandings of death and, especially, the practice of dying, albeit always accompanied by a sense of regret, even anguish, as Professor Nigel Biggar notes.[25]

It might be objected that my suggested controls are excessively severe. But for those very reasons that I have been at pains to explore, such a restrictive code is surely necessary. It would counter the wanton deceitfulness and wilful disregard that cannot but obtain, particularly – it must be stressed – *over time*, once society and especially those directly involved in its practice, become attuned to the provisions of any legal reform. The following passage by Sissilla Bok, exemplifies precisely what I have outlined above, referring both to legal and medical practice:

> Intellectuals are more prone than practitioners in most lines of work to fall into the trap of believing that they can devise a set of rules to guide participants in an inherently dangerous practice. They may take for granted that the *rules – once enacted into law – will work in the intended fashion*. Too often, experience shows that their *faith* turns out to be *misplaced*. At times, those to whom the rules are supposed to apply *don't know of their existence*. At other times, the rules are known, but *people do not obey them*, or *evade them* by finding loopholes or *ambiguities to avoid compliance*. [26] [My emphases]

My conclusion, therefore, is that there is still great uncertainty and incoherence attending the practical aspects of AS for which Lord Falconer presses. That conclusion is one which all Christians need to be aware of – so to confidently articulate those itemised shortcomings in future debate, and to realise that a change to the Suicide Act (1961), specifically, is unnecessary. Falconer has steamrollered his campaign for legal change over many years – and it is seemingly impregnable. Here, from an essentially clinical perspective, I have questioned some of the central planks of his proposals, and shown them to be in need of modification – as well as capable of being subsumed into other possibilities.

* * *

1. A type of brainstem stroke which leaves the subject severely impaired. Nicklinson was a 'front' for Dignity in Dying, an organisation whose aim is to change the law on assisted suicide. Nicklinson died suddenly from pneumonia, which denied him opportunity to retain total control over his hour of departing, and to witness the legal enactment he so desperately sought.

2. London: Demos 2012. *The Current Legal Status of Assisted Dying is Inadequate & Incoherent.*

3. Badham P, 2009. *Is there a Christian Case for Assisted Dying?* London: SPCK; Pitcher G, 2010. *A Time to Live: The Case against Euthanasia & Assisted Dying.* Oxford: Monarch Books.

4. Marsh MN, 2014. The Debate On Assisted Dying (AD). The Church Times, 28th March 16.

5. Switzerland is the only country that allows suicide tourism.

6. Canon Dr. James Woodward (S George's Chapel, Windsor) was a notable dissenter. London: Demos Report, 2012, Appendix 3: 'Statement by the Reverend Canon Dr James Woodward', (on why he disagreed with the Report's more specific conclusions – re Chapters 11 and 12).

7. Assisted Dying Bill [HL Bill 24]: London: The Stationery Office, 15 May 2013.

8. By 'physically', I mean a person – despite whatever illness

causes the desire for an end to life – to be capable of administering the lethal dose without external aid.

9. Bok S, 1998. 'There is every reason to look with wary eyes at any calls to put much faith in the power of 'appropriate legal safeguards'. In: Dworkin G, Frey RG, and Bok S (eds), *Euthanasia and Physician-Assisted Suicide.* Cambridge: Cambridge University Press 135.

10. We just need to recall those corporations intent on evading taxation; or those who make their living outside the law, involving drugs, prostitution, people trafficking; gangs involved in child prostitution; and those concerning themselves with internet child pornography, etc.

11. A-E: A: the continued pregnancy would involve a greater risk to the woman's life; B: termination necessary to prevent grave permanent injury to her physical /mental health; C: not exceeding 24 weeks, continued pregnancy would involve greater risk to her physical/mental health; D: as C but additionally causing greater physical or mental injury to her existing children; E: if child born, substantial risk of serious handicap due to physical or mental abnormalities.

12. London: The Office for National Statistics (ONS).

13. I am grateful to Dr. Jon Martin, S. Joseph's Hospice, London, for that insight.

14. This was intentional, so as to prevent doctors from being taken to court. That continued inclusion to the Act (1967) was based on a 'famous' London gynaecologist who performed an abortion following rape, but was subsequently acquitted, Rex versus Bourne [1939] 1 K.B. 687.

15. Concerns over the fitness of doctors in their professional roles have been voiced previously. See Bok S, 1993. Impaired physicians: What patients should know. *Cambridge Quarterly of Healthcare Ethics* 2: 331-340.

16. Baroness Ilora Finlay, Hansard HL Deb, Vol. 712, Column 607, 7th July, 2009.

17. The now infamous case of Al Megrahi, the supposed Lockerbie bomber, is a case in point. Although assessed by medical personnel to have less than three months to live, he was still alive more than 2 years after being released from a Scottish prison. We might all consider the case of Nelson Mandela. It is obvious to everyone that he was on a steep end-of-life trajectory, but who among us could have prophesied when he would die? In fact, he lingered for longer than most of us expected.

18. Figueiredo J & Frank J, 1982. Subjective incompetence: the clinical hallmark of demoralisation. *Compr Psychiatr* 23: 353-63.

19. Kissane D & Kelly B, 2000. Demoralisation, depression and desire for death: problems with the Dutch guidelines for euthanasia of the mentally ill. *Austr NZ J Psychiatr* 34: 325-33: Clarke D & Kissane D, 2002. Demoralisation: its phenomenology and importance. *Austr N Z J Psychiatr* 36: 733-42.

20. Chochinov H, Wilson K & Enns M, 1998. Depression. hopelessness, and suicidal ideation in the terminally ill. *Psychosomatics* 39: 366-70: Beck A, Kovacs M & Weissman A, 1975. Hopelessness and Suicidal Behaviour: An Overview. *J Am Med Assoc* 234: 1146-9.

21. Hollins, Baroness, Hansard HL Assisted Dying Bill (Committee), Column 1920, 7th November 2014.

22. Grey-Thompson, Baroness, Hansard HL Assisted Dying Bill (Committee), Columns 1924-5, 7th November 2014.

23. Swinfen, Lord, Hansard HL Assisted Dying Bill (Committee), Columns 1926-1928, 7th November 2014.

24. Badham, *Christian Case* 123.

25. Biggar N, 2004. *Aiming to Kill: The Ethics of Suicide and Euthanasia.* Cleveland: Pilgrim Press 114.

26. Dworkin, Grey and Bok, *Euthanasia* 135.

Chapter 11

Assisted Death (AD): Theological Proposals

The likelihood of legalised AD (I use this terminology purposely, in order to remove myself from Lord Falconer and his Assisted Suicide movement) looms large in the public sphere. In the preceding chapter, I discussed a legal approach to this problem, based on the wisdom that has accrued to us through the working lifetime of the UK Abortion Act 1967. Here, I want to proceed by thinking about theological approaches to AD, an aspect of the confrontation which has not well been articulated, in my view.

I have several reasons for wanting to adumbrate a theology of assisted dying.

First, many Christians in their heart-of-hearts probably espouse AD. Nevertheless, when confusedly agonising over the long, drawn-out demise of precious relatives whose degenerate bodies prevent end-of-life fulfilment, many might find consolation in some relevant theological reflections.

Second, if relatives continue to harbour deep feelings of guilt after having been caught up in such heart-rending decisions, a framework for assuaging that guilt might possibly be welcomed.

Third, many believers (irrespective of immediate looming death) might consider AD anti-Christian.

Fourth, ethical literature on AD and cognate matters is extensive and daunting. In failing to provide definite answers, it offers scant help for the average person in mounting effective responses or challenges.

Next, the AD 'debate' has pitched pro-lifers against the euthanasia lobby, again without much usefully focussed outcome.

Finally, some may conclude that church and academy have not articulated a coherent AD theology that is of practical use in these difficult circumstances. It is also important to note that my use of AD applies generally to whom this end-of-life strategy might be considered appropriate, irrespective of physical or mental competence, age, mobility, and so on.

My quest is simple: if (or when) UK law ultimately sanctions AD, could Christian theology support it? However, before thinking about that, it is necessary to consider perceptions of death throughout 21st-Century Western society; next to evaluate the conventional sentiments brought to bear against AD, and third, to reconsider current understandings of the meaning of 'person' in respect of AD.

Death in Western Society

We shall all die. In the West, however, we have forgotten how to approach death, and few people ever confront a dead body. In passing, this is a subject that needs reviving.[1] Yet despite that, death and dying, paradoxically, are constantly before our eyes whether portrayed as television drama, murder on city streets, the outcome of war, or from major headlines which, in recent years and of increasing concern to many, highlight both the degrading abuse of the elderly and frail in hospitals or care homes, and the attempted changes to the law governing assisted dying. Although the individual may be almost dead, some parts of society seem to persist in refusing the facility of a welcome and peaceful death withdrawn from the trappings of clinical support.

Society thus endures ambivalent attitudes towards death and dying, perceiving it to be inevitable yet fearful; terrible and life devastating. The resulting ignorance creates further problems, especially for those with a religious faith, resulting in added uncertainties about resuscitation procedures. These are widely seen in many instances as unnecessarily prolonging any useful life for those smitten with terminal forms of disease or dysfunction.

However, a multiplicity of reasons underlie the growing

trends supportive of legal change of which we must be aware: the progressive loss of family cohesiveness and its tradition of care for aged relatives; the secularisation of society which sees death as of no intrinsic significance; increasingly malign attitudes towards the physically or mentally weak; the persistent media 'drip-feed' that the aged are useless, drain limited public resources, and need elimination; the pursuit of everlasting youthfulness and beauty (hormones, corrective 'plastic' surgery, anti-senescence potions); and the overriding assertion that 'personal autonomy' and 'control' are key factors in end-of-life choices over one's existence here on earth.

These remarks indicate the uncertainty, confusion, and ambiguity surrounding death and dying: these are subjects needing to be understood if assisted dying is to be part of daily commerce. The legality of assisted dying has been affirmed in certain countries, is being considered by other jurisdictions, while some form of legalised assisted dying will inevitably involve the UK in the coming years.

From an (exclusively) Christian perspective, what does death mean? Job pertinently asked: 'If a man die – shall he live again?' (Job 14: 14). Christians assert a credal belief in 'life everlasting', affirming, additionally 'the resurrection of the dead'. But does that mean, or even imply, that we become either immortal, or that we somehow become a part of eternity – or, perhaps more properly, enter into the eternal life of the Godhead? In thinking about death and dying, we have to consider among other things, the role of bereavement and grief following the decease of a loved one.

In her recent book, the popular British TV presenter Sally Magnusson describes the long descent into end-stage dementia undergone by her mother.[2] In the final chapters of the book, Magnusson ponders questions about personal identity, consciousness, and the brain – in other words, she raises queries about who and what we are – and, most interestingly, what

209

remains (if anything) at the very end. Another relevant, contemporary example involved the BBC financial reporter, Robert Peston.[3] Having recently lost a very dear wife from breast cancer, he was devastated and experienced a considerable sense of extreme loss, loneliness and darkness. The pain of this bewildering emptiness, even though expected, was to some extent overcome through the influence of his two sons, and bereavement counselling. But he also became acutely aware that at work or social gatherings, death and its subsequent distress were never spoken about. That is, there was no apparent external perception by those outside the confines of family that a recent death is, or even could be, an existential source of distress.

Unlike the Victorian era where heavy mourning clothing was acceptable, followed earlier in the last century by the wearing of a black armband or necktie, today society prefers not to be reminded. While society lives with this separation from death, it also encounters a surfeit of death. And yet despite being surrounded by death and all its horrors, people have no idea how to deal with bereavement, what to say, or how to cope with it when they are alone with the bereaved individual, thus experiencing acute embarrassment when so confronted.

Death, or dying, is a subject waived aside, air-brushed from public consciousness. Irrespective of these difficulties (and relevant to optimal outcomes in individual cases), when death finally comes – at home or more commonly in hospital – the body is rapidly whisked away by undertakers who magic the corpse to the grave as quickly and unobtrusively as possible.

'Christian' Arguments Brought Against AD

In proceeding, we need to remove those religious themes which, to my mind, constitute irrelevancies when considering the theology of assisted deaths, or euthanasia in general.

Jesus' mission was one of compassion towards those for whom the Kingdom was inaugurated, and the basis of what we know as the Golden Rule, and similarly espoused by the Oxford philosopher GJ Warnock whose views we have already encountered.[4] Thus, for Warnock, *sufferability* encapsulates the human predicament, and that, crucially and ultimately, must embrace taking care of those unable to speak up or even defend themselves against the undesirable edicts and actions of others. While the Golden Rule offers a sound basis for treating everyone as would be desirably expected in return, we need to critically examine those commoner traditional defences which are hauled to the ramparts by Christians when assisted dying or euthanasia threaten.

(i) Commandments: these represent a framework governing the behaviour of the Eucharistic community, gross deviations from which impair the ecclesial union with Christ (1 Cor 6:9-10). Paul's emphasis in his letter dwells on right social embodiment, so that the selective (and legally controlled) assisted suicide or death of individuals in great distress (1 Jn 3:17) could hardly be deemed as outright 'murder'. Even the cold-blooded murderer Cain is protected from the vengeance of the vendetta – and its multiplicity of ensuing effects.[5]

In the biblical canon, the Hebrew command in the Sixth Commandment, 'Do not kill' (in the Hebrew, *lo' tirzaḥ*), widely encompasses murder in hot and cold blood, military executions and slaughter of animals, transferred into LXX (*ou/me* plus *phoneuo*) and the synoptic gospels as prohibition against murder, rather than manslaughter. But the commandment, as such, cannot be fixed to specific legal examples, and instead prohibits all wilful or lawless transgressions against human life or which, in aggressive hatred, infringe the other's right to life, contrary to the good of the community and its consequent irreconcilability with human dignity.

Clearly these direct (deontological) commands fail to cover every exigency: if they did, there would be far less need to engage in so many heart-rending 'bioethical' debates. On these grounds, it could hardly be asserted that the Sixth Commandment provides warrant for arguing against the possibility, and selectivity, of assisted dying.

In passing we should notice that the commandment is often tied to the instruction 'but need'st not strive officiously to keep alive'. This often seems to be honoured as a text of ancient, profound wisdom to which a solemn due regard should be paid, particularly in the context of assisted dying. However, it is nothing of the sort, and should be disregarded with as much cynicism as it was written.[6]

(ii) Sanctity of the flesh (or body): this is one of those vague phrases which is often employed as a canonical epithet, stifling ethical debate and, (like 'commandment'), failing to find a niche in many 'Christian Dictionaries'. Even Glanville Williams' book is a misnomer embodying a consistent polemic against ecclesial directives concerning death.[7] Sanctity of the body, I think, differs from personal *sanctification*, that is, a making holy ('theosis') through sacramentally derived divine grace – received through baptism, and eucharistically. We must not be confused about differences between the spiritually-endowed *body* (persona: hupostasis) viewed as holy vessel (1Cor 6:19) from mere bodily *flesh* (substantia: ousia).

Indeed, if we think clinically, then the flesh (qua body) has been subject to much manipulation by the medical fraternity. If we could imagine one human bodily cartoon displaying all the metal and plastic replacements, technological gadgetry, applied pharmaceutics, and tissue transplants that are in use these days (with more, especially at macromolecular and genetic level almost certainly in the pipeline), there is not much of the flesh which has not succumbed to this steady encroachment – and, we must assert – for the benefit of humankind. Even the Godhead might applaud these revolutionary advances: after all, it reflects actions partly through which the inauguration of the Kingdom was manifested in Jesus' earthly ministry (Isa 61:1-2; Lk 4:17-21).

Indeed, in our own times, we are beginning to witness to the blind seeing, the deaf hearing, and the lame walking – although contemporary medical science does not bring it off with quite the panache and miraculous sleight of hand demonstrated by Jesus.

(iii) Made in God's image: this assertion invariably stalls informed discussion. Christians harbouring this idyllic vision would be brought up sharp when challenged with the existence of brain-damaged children, or the severely demented – that is, the types of human being who might possibly be considered for assisted suicides or killings, and towards whom this paper is directed.

In attempting to solve the meaning of 'image', James Barr was non-committal (given its stern overshadowing by Deut 4:13, 15-16) while Crouch has used parallels between the parentage of Adam-Seth (Gen 5:2-3), and humanity with God, drawing on themes of createdness and care (Isa. 43:1, 5-7; Jer. 3:12; 31:22; Mal 1:6, 10; Prov. 2:11-12).[8,9] Thus, as we are all 'children of God', we should respect and afford dignity to each other as brothers and sisters. Yet the call to be God's vice-regents (Gen. 1:28b) has hardly fostered an earthbound Eden. As we learned above, our inhumanity to each other continues, making us shudder at the levels of depravity to which the human race throughout history has sunk in treating others with such atrocious disregard.

Despite such explanatory (and diverse) attempts, the 'image' problem is resolvable when envisioning Jesus who, manifesting the perfect hypostasis in his humanity, signified the true Image of God. Taken together, it would follow that if every human being could aspire to be truly in God's image, then love of one's neighbour and a mature approach to the care of the world and its resources would inevitably occur. Despite that, this percept offers scant help in modulating any Christian-based strategy for assisted dying.

(iv) 'The Lord gives and the Lord takes': in continuation of these themes, another 'protective' argument needs examining, in

its implication that we have no right whatsoever, as Christians, to take the life of another individual, not even in advanced clinically deteriorated circumstances – it rests with God himself. The problem with the uncritical use of biblical quotes is that they can often be overridden by other antithetical positions.

So to return: does God give and take? If we review the 20th Century, more people have been killed or exterminated than in any other previously documented era. This has not only involved all the losses of the two World Wars, but considerable 'ethnic' and political cleansing, tribal extinctions, and the results of sometimes massive starvation in Africa. Throughout history, large swathes of population have been decimated through epidemics – the bubonic plague, puerperal sepsis, tuberculosis, more recently AIDS, and now the acute Ebola-induced haemor-rhagic fever. These well-known facts need no further elaboration.

The point, clearly, is that one could hardly allege that these disastrous eventualities are engineered by God in order to dispose of humankind: that would be a ridiculous conclusion. Other than geophysical catastrophes and massive infectious epidemics, the rest have been largely unleashed by mankind against his fellow man. So, again, the weakness of the appeal that 'God takes', specifically if argued in favour of assisted dying, is all too evident.

(v) Gift and Being: or Being Done-With?: in returning to the idea of life as God's given gift, we should further remember that we get life from elements already living – ovum and sperm, representative of living matter evolved in continuity from its first origins on earth. What we do get from each act of human creation is a *being*, and from which, the opportunity for true Personhood. It is all too easy to idealise concepts of life as a precious God-given gift wrapped in attractive paper, tied with a beautiful bow ribbon, and hence inalienable or non-returnable. But that kind of talk needs further purging.

Life is hard and gritty rendering us all vulnerable to the

impact of McCord Adams' 'horrendous evil or horrors' – destructive of all true 'meaning-making-in-life' for any human being at whatever age – a mother's stillbirth, parents of a 6-month-old baby dying of acute leukaemia, a child developing muscular dystrophy and becoming wheelchair-bound for the duration, a teenage undergraduate reduced to low-level mental capacities through viral Herpes and simplex infection of the temporal lobes, or a young male bread-winner accidentally killed through health and safety workplace violations.[10] Neither might we find it too appealing to have our bodies non-functional, resulting from progressive neurological degeneration or eaten away by metastatic cancer – as though worms were steadily dissolving our flesh and destroying the fabric of our very existence. McCord Adams determines these 'horrendous evils' as non-explicable in terms, necessarily, of rampant sinfulness, wayward freewill options, or moral issues, but to God – since he created the world into which we enter and are expected to grow – despite many horror-inducing vicissitudes of divine creation engendered through our freedom, and not only from man's inhumanity to man.[11] Second, even the earth as a living planet, so geophysicists inform us, must undergo tornados, volcanic eruptions, earthquakes, tsunamis, and monsoons – irrespective of the considerable loss of life, livelihood, community, and economies thereby exacted. Third, we are directly exposed to mutational gene disorders which debase and rob us of life.

And 'Gift'? From all this, it would not be unreasonable from a Christian perspective, given the particularity of a life-degraded body, to return it as 'no longer fit (so the phrase goes) for purpose', and hence as radically inimical to any possible hope of further meaning-making-in-life. And this would apply whether life was regarded as gift, or, as others suggest a loan – although in this type of case, a rather dodgy loan. Moreover, God as Father has realised the human predicament – not merely of *sin-*

redemption, but of *lost-meaning-in-life* redress from horrendous evil by assuming human flesh through the Son, and experiencing for God-self the very exigencies of being en-fleshed, through the actuality of incarnation.[12] Stage one horror relief, as Adams defines this first phase in our redemption, recognises the presence of God, as incarnate Son, to be the very proof of that being-with-us-in-adversity: God with us (or in the Hebrew: Im-anu: el).

These thoughts go a long way in clearing away all the foregoing approaches that, in my view, scarcely meet the issue at hand. Thereby, the way is opened up for further engagements with AD. Indeed, we now have a blank slate! Moreover, in moving forward, it is essential for a moment to recall the sentiments written with regard to our considerations about the theological aspects of personhood (Chapter 6). There, I reiterated the problems associated with creatureliness, of having to live one's life in a created universe.

There, to briefly recap, we were reminded that we cannot frame any persons in their entirety by use of earthly labels: that is, we can neither fully know ourselves, nor others – even those especially close to us, nor they us. In addition, we are so frustrated by being created that we cannot transcend our need to escape the bounds of enfleshment: that is, to become divine. Many try to escape this dilemma by achieving so-called happiness through money, wealth, social standing, and trying to love, but within the confines of corporeal existence. But that will not do. Our yearning for true knowledge and understanding is to become 'divinised' through baptismal union with the Godhead, thereby becoming His 'children by adoption', and thereby achieving complete 'freedom'. Those words only acquire their full meaning when interpreted metaphysically within the Godhead, and not physically through worldly appropriations. Indeed, to insist on the latter would be ridiculous: for example, there is no such quantity as complete freedom within this world. That ultimate acquisition of freedom from our unfulfilled yearnings and hence 'rest' (stasis) within the Godhead involves dying, a process made tolerable because Jesus has not only died for us – but more importantly, risen.[13] That is the Christian guarantee of hope, thus making the 'sting' of death not only bearable, but tolerable as well.

But I realise there is a problem here, in talking solely about baptismal 'becoming' – in, and through – the death of Christ. However, when these words were being written, everyone paid allegiance to some kind of god. The message of Christianity was the need for baptism in order to become a member of the fold, in communion with all other baptised people of God. There was no other alternative, and that is why I am working with the theological concept of being baptised into the Godhead as the (only) way possible for us to escape death and live a life with those who have already passed on and are already members of the Communion of Saints.

Having made that point, we can proceed with the foregoing sentiments in mind to consider theological approaches to AD. I suggest we can derive relevant principle from Jesus – his life; passion and death; and ultimate resurrection and ascension to God as our heavenly, adopted Father.

A Theological Framework Incorporating AD

(a) Kingdom ethics and loving kindness: in inaugurating the Kingdom of God, Jesus revealed the 'compassion' necessarily and unreservedly required for the weak and diseased – the halt, blind, and deaf – an attitude notably inconsistent with current definitions of 'persons'. Moreover, the gospel's 'splagchnizomai' demands more robust a translation than mere compassionate sympathising – it is more a sickening, pained anguish even tempered, maybe, with a flash of anger. But is compassion all that is needed here, rather than empathy – which, precisely defined, conditions a dispassionate, accurately judged assessment which would be deemed essential for effective

decision-making for anyone *in extremis*.[14]

With that strategy, we avoid the egocentric bias incurred by misattributions of the victim's predicted needs, that is, as risked by merely demonstrating 'compassion' or 'sympathy'. In passing, we should notice that the Good Samaritan's well-organised *actions* (Lk 10.30), following an initially perceived compassion (from the Greek *èsplagchnísthē*), were decidedly *empathic*. This distinction is vitally important in dealing with end-of-life issues, especially if the subject is helpless. AD should be procured disinterestedly with the *person's* needs in mind, not the observer's.

Easily said, perhaps, but incredibly difficult when *I* cannot fully know *your* intent, nor *you* in your deepest desires, particularly when the loneliness of death threatens.

In continuing, we should note that Jesus certainly was fully aware of environmental exposure to 'horrendous evils' – man's inhumanity to man, life-wrecking geophysical catastrophe, the inheritance of mutationally-damaged genes – but nevertheless, all embodied outcomes of the worldly habitation God created for us.[15,16] Comparatively for Jesus, we might usefully note, much of this was of lesser concern (beheading, Lk 9.9; lost body parts, Mt 5.29-30: 18.8; killing, Mt 10.28). For Him, The Kingdom, for those 're-born baptismally from above', required achieving justice for and love of our neighbour (from the Hebrew, *hesed we'emet* and the Greek, *charis kai aletheia*) – two themes constituting a noteworthy thread (forty-two occasions) throughout the Hebrew and Greek scriptures.

The ontological anthropology, 'being-in-Christ', of being partly 'en-hypostasised' baptismally into the Godhead as the non-definable basis of temporary earthbound 'who-ness', suggests that discarding a rotten body – and especially when death threatens – now becomes relatively unimportant. Our time-limited histories have since already ascended and been remembered within Godhead: 'Your prayers and acts of charity have gone up to heaven to speak for you before God' (Acts 10.4).

Kingdom ethics in our present day embodies, even warrants, mobilisation of compassionate/empathic strategies for any loved-ones suffering terminally-irreparable organ or system failure, giving relief in these extreme circumstances, and, where thought necessary, helping them onwards to the Godhead in whom entitlement of placement is guaranteed – baptismally and eucharistically. And that help would be given dispassionately by those who really cared for whoever they were trying to deliver that merciful kindness to in end-of-life situations.

(b) Waiting: end-of-life trajectories invariably require waiting, as admirably summarised by Canon William Vanstone.[17] He observed how all of us wait, in being exasperated with others' failure to complete their duties, power cuts, computer break-downs, road blocks, or travel-interrupting weather.

Jesus, having been the man of action in actively dispensing God's power (*exousia*) was made to wait passively after arrest, thence to be done unto. As Vanstone notes, all the Greek aorists in the gospels suddenly turn from active to passive mood. As the Man of Sorrows – beaten, bleeding and scourged – he truly represents all those *in extremis*. Terminal disabilities or illnesses radically curtail personal activity thus enforcing, through the inevitability of lost control, a forfeiture of hitherto-enjoyed autonomy. Crucially, and paradoxically, our bodies at that stage have scant usefulness relative to a carefree past when self-enacted agency, demeanour, and outreach prevailed. The enforced result entails a necessary and inevitable reluctance in yielding to the imposed waiting *for*, being waited *upon*, and resignation to 'the system' and its patterns of service delivery.

Those who insist on retaining 'personal autonomy', remaining 'in control' and desiring to 'determine their hour of death' are essentially deluding themselves, since coming under the care, even supervisory management of others, inevitably means losing personal overall control which they erroneously imagine being able to hold on to, and to continue exerting.

End-of-life matters incur much waiting once bodies or minds devalue fulfilled living. The point is that having to wait, in affecting so many elderly people, is a holy state as reflective of Jesus' Passion. Our own passivity in the face of subjection to others' derivative powers is reminiscent of His interrogation and humiliation before Pilate and the Chief Priest.

(c) Sacrifice: in the ambiguity of life, our human earthly existence is always fully open to ruin by the kind of creation God made for us to inhabit, and yet God-as-Father alone is (or should be) able to save us from all ultimate degradation and despair, although not necessarily in this life. An individual's choice of death through legally sanctioned AD permits release from grossly dysfunctional bodies obstructing continued effective *meaning-making-in-life*.[18] Here the platitudinous idea of the non-returnable, divine gift of life could be questioned, especially in the presence of profound life-destructive bodily failures.

Religious practice is the locus of rituals. Christianity, although never condoning Temple-type offerings, retained the central premise of Jesus' once-for-all 'sacrifice' on the cross. We, the Christian community, own nothing and offer what is available: 'Of your own do we give you'. Thus, in the presence of end-stage bodily dysfunction, which is destructive of people's lives, we, as its members and as individuals, should consider AD as a means of sacrificing a completely useless body back to the Godhead. Aligning the death of 'end-stage' patients to Jesus' sacrificial death makes additional sense, as both reflect physical and existential suffering. But, through Jesus' sacrifice, human sufferings are sublimated, sanctified, made holy: 'we offer our ... bodies ... to be a living sacrifice'. Added meaning to a Eucharistic celebration, involving all concerned, would thereby be realised at the appropriate moment of farewell as has been rightly suggested by Professor Paul Badham.[19]

(d) Would AD invoke God's anger?: would God find displeasure with AD in these complex circumstances? Answers

here would help to assuage the guilt of relatives ensnared in such heart-rending dilemmas. The answer (Eucharistic Prayer) is 'When they turned away and rebelled, your Love remained steadfast'. The key here is divine love faced by outright rebellion.

God's supreme incarnational act of redemptive love was repulsed by mankind's return of Jesus as a mangled, crucified body bearing the notice 'No thanks'. In AD, some victims, likewise, have the overwhelming urge to return their bodies with a note saying: 'This flesh can no longer support the life I should wish to lead'; the body is offered up and returned.[20, 21]

Yet, in overriding mankind's arrogant rejection of the Son given as Redeemer of the World, God nevertheless triumphantly resurrected Jesus' body as sure, consistent demonstration of his ever-present, steadfast love for humankind and, indeed, the world of disasters arising from the habitat he created for us.

That is a love beyond all human comprehension, in which we believe, to which we bow the knee, and fervently proclaim will never yield to the grossest deviations within creation. From that, all faithful believers may derive satisfaction that in giving a charitable ending to those few individuals suffering severe grievous bodily distress, God's anger would not be invoked. The framework elucidated above, based upon the life, passion, and sacrificial death of Jesus, should help to dispel any reticence in proceeding in those specific cases where AD is requested or considered appropriate to circumstances.

'Lord, now lettest thou thy servant depart in peace'.

Postscript

I have attempted to offer some thoughts to those who either feel AD is wrong, or experience deep feelings of guilt over this matter. Decisions on these issues are most difficult, even impossible, especially since we do not know ourselves when acting contingently on available short-term information for, and with, any victim (Chapter 6). The latter person may, additionally, be psychologically uncertain, or suffering agonising loneliness, demoralisation, or hopelessness as a result of bodily disintegration (Chapter 8).

AD involves babies, infants, adolescents – so growing old or living longer is not the sole consideration here. Life (or 'apparent' life) can be prolonged for no good reason. In such heart-rending circumstances AD may appear to be the best option for terminating a life beset by persisting unrelieved suffering. Such acts require empathic appraisal, rather than primarily compassion or sympathy.

The thoughts outlined here are based on the life, passion, and sacrificial death of Jesus, with whose three phases of life on earth may be aligned the spectre of human suffering, and which thereby is made holy. It is hoped that the view proposed offers comfort to those wrestling with these perplexing decisions.

I firmly espouse the Orthodox view of baptism as means of becoming 'known' to the Godhead in whom there is temporary lodging as a new 'hypostasis', and which perpetuates that metaphysical 'rebirth from above' away from a creaturely life, through death and into the beyond.

Thereby body or mind (brain) following irretrievable degeneration can be dispensed with, in the final hope of putting on immortality.

* * *

1. Paul K, 2015. The Ars Moriendi: A Practical Approach to Dying. *Modern Believing* 56: 207-220.
2. Magnusson S, 2014. *Where Memories Go: Why Dementia Changes Everything*. London: Two Roads (Harcourt Brace).
3. Gibbons K, 2014. How Robert Peston coped with his 'Overwhelming Loss'. London: *The Times*, 1st February 36.
4. Warnock, Object, 151.
5. Gunton C, 2002. *The Christian Faith*. Oxford: Blackwell 148.

6. The entire block comes from 'The Last Decalogue', a poem written during the 19th Century by Arthur Hugh Clough (1819-1861). It was a cynical attack on each commandment following his loss of faith, and with it, a teaching Fellowship at Oriel College, University of Oxford. We need attach no importance to this often mis-represented, and mis-quoted, epithet.

7. Williams Glanville, 1958. *The Sanctity of Life & The Criminal Law*. London: Faber & Faber.

8. Barr J, 1968/9. The image of God in the book of Genesis – a study of terminology. *Bulletin of the John Rylands University Library [Manchester]* 51: 11-26.

9. Crouch C, 2010. Genesis 1:26-7 as a statement of humanity's divine parentage. *Journal of Theological Studies* 61:1-15.

10. Adams MM, 1999. *Horrendous Evils And The Goodness of God*. Ithaca: Cornell University Press 174: Young F, 1982. *Can These Dry Bones Live?* London: SCM Press 56ff.

11. Young, Dry Bones 56ff.

12. Young, Dry Bones, 36; 41.

13. Zizioulas J, 1993. *Being In Communion*. New York: S. Vladimir's Seminary Press 105-6.

14. Marsh MN, 2013. Empathy: Mirroring Another's Predicament – Or More? *Antonianum* 3: 409-430.

15. Adams, Evils 174;

16. Young, Dry Bones 1984 57-9.

17. Vanstone WH, 1982. *The Stature Of Waiting*. London: Darton, Longman &Todd.

18. Adams, Evils 26-28; 174.

19. Badham P, 2009. *Is There A Christian Case for Assisted Dying?* London: SPCK 123.

20. Ashby G, 1988. *Sacrifice: Its Nature And Purpose*. London: SCM Press 124-131.

21. Adams MM, 2006. *Christ And Horrors: The Coherence Of Christology*. Cambridge: Cambridge University Press 270-281.

Part V

Resume

Chapter 12

On Being Human

...Six months from now her baby would be born. Something that had been a single cell, a cluster of cells, a little sac of tissue, a kind of worm, a potential fish with gills, stirred in her womb and would, one day, become a man – a grown man, suffering and enjoying, loving and hating, thinking, remembering, imagining...What had been a kind of fish within her would create and, having created, would become the battleground of disputing good and evil ...What had blindly lived within her as a parasitic worm would look at the stars, would listen to music, would read poetry... would become a human body, a human mind. The astounding process of creation was going on within her...[1]

In this evocative vignette reflecting Haeckel's (1834-1919) now defunct dictum 'Ontogeny recapitulates Phylogeny', the novelist Aldous Huxley weaves, through textual antithesis, the threads of biological determinism against those of social programming and environmental influence.[2] There is a considerable difference between this romanticised picture and the evolutionary scheme offered by his brother, the noted biologist Sir Julian Huxley, encountered above (A Theological Anthropology: Chapter 6), but it is relevant to our thoughts about humankind.

Their respective accounts, nonetheless, highlight the marked contrasts between biologically ordered intra-uterine gestation and post-natal inculturation, the latter vital for lives capable of being developed and fulfilled at their highest cognitive and artistic levels. Nevertheless, both lack reference to the loci upon which the acquisition of full personhood is usually predicated – a body, a brain, but also a community and within it, a relational stance with others, through whom those attributes are most effectively developed and expanded.

There are numerous books devoted to the subject of humanity, but I think mine is different in pointing to aspects of living the life of a human being which rarely, if ever, undergo *in depth* analysis in other texts. In being concerned with the business of personhood, with being human, or what it is like having to live as a human person on this earth, I have divided my approaches into four sections, entitled Distinctiveness, Dignity, Disability, and Disposal. These headings encapsulate key reference points on which such approaches should surely hinge, but they are not definitive, being subsumed within my overriding theme of 'Who-ness', pointing to the true person in comparison with 'What-ness' which, in defining worldly features which may often be shared with others, fails to encapsulate personhood in its entirety, and certainly not in its singular uniqueness.

The true definition of any person, as noted in my initial chapters (Chapters 1-6), cannot be derived solely from physical attributes, as encapsulated here under Distinctiveness (embodying physical anthropology; genes; consciousness; language). Moreover, many other accounts tend to favour the loftiest and, not withstanding, the most abstract features of human personhood – 'life lived at its highest cognitive, artistic and emotive levels'. It would seem that being disabled (Chapter 8: On Being Disabled, Dysfunctional, & Disfigured), or even a glimmer of awareness about everyone's gradual descent through natural degenerative change or progressive disease towards an inevitable death, and the accompanying process of dying – whether short or drawn-out (as discussed under Disposal), are rarely deemed part of the process of living as 'normal' human beings. Neither do these features receive adequately detailed exposure when humanity is being discussed, since our cognitive abilities are frequently held up as an exemplar of supreme human excellence. But, regrettably, this approach fails to take notice of the wide variations in cognitive prowess definitive of the human species, dwindling to almost nothing for brain-damaged people. Yet the latter are not to be branded as non-human things, despite a widespread revulsion throughout society towards them and their misfortunes (Chapter 8). Therefore, we do need to keep in mind what sort of difficult lives they often have to lead and to be sensitive to the predicament thereby engendered.

Neither is there anything specifically distinctive, nor 'unique' in our genetic make-up (Chapter 3: Genes), since our total gene pool ('genomes') is invariably recycled from what has gone before. Even socially, doing the same work robs people of the guarantee of 'uniqueness' while one's name, so often reduplicated by others, no longer serves as a guarantor of self-identity.

But having stated all that, we have cause to consider what other options are available (if any) which provide sufficient grounds for defining a human being. Problems in defining human beings still continue to exercise theological, philosophical, and scientific minds: 'It is', as Paul Tillich argued, 'the ultimate question'.[3] Even now, then, it seems as if we have not quite grasped this matter in its entirety: how do we frame a person?

Framing the Person

In starting, I refer to two books which have made defining being human their subject matter, van Huyssteen's *Alone in the World?* and Green's *Body, Soul, and Human Life*.[4,5] Green, rightly so, is less interested in genes, consciousness, even the possession of a 'soul': his critical point of departure is the Genesis account of creation displayed on the first page of the Bible – mankind-in-God's-image (his book, p61). He therefore concludes that while science and the biblical account both engender notions of human embodiedness and relationality, the biblical command demands further relationship and orientation with the Godhead (his book, p71). For him, 'image' suggests a relationship both diachronically with the animals and back through creation, but also synchronically in relationship towards others, and the Other.

In his book, based on the Gifford Lectures given in Edinburgh in 2004, van Huyssteen asserts (likewise) that humankind is steeped in the theological tradition of the *imago Dei*, yielding to the belief that all humans are different and distinct because of that biblical or divine relationship.

Both he and Green fail to define the 'image' epithet. Conversely, those who have tried to discover its true meaning will have soon become overburdened, if not a little disillusioned, by the vast bundle of relevant papers written, and their inconclusive outcomes. In his 'inter-disciplinary approach towards human 'uniqueness' as comparatively assessed through science and theology', van Huysseen persists in equating uniqueness with distinctiveness, an approach which in my view is mistaken. One can be extraordinarily distinctive (a florid growth of ginger hair, certain clothing, or some other striking feature or possession, even one's name) and yet such features can also belong to another. While such types of marker would certainly be distinctive they would not be, nor could they be, unique. Or, to put it more tersely, the Chappell twins from Philadelphia are conjoined at the head. They are certainly distinctive, but since several other (stage 4) twinning outcomes produce the same anomaly, it is not unique.[6] Hence, these terms of attribution should not be elided.

I have other feelings of unease towards these books.

To continue – I resist the 'image of God' theme: I think it is the wrong model to apply in formulating the basis of a true anthropology of being human, given that no one really knows what it actually means.[7] As a result, authors, having drawn a blank, usually proceed by then imposing their own interpretations on their readership. That is a move hardly likely to gain support or elucidate the problem. The objections that I pose against the image-of-God option, however, gain far greater traction when viewed against its use as a major prohibition against assisted dying. This approach most critically impinges on 'person' definitions when viewed especially from end-of-life perspectives. It merely becomes a conversation stopper, rather than being logically developed into a persuasive argument (and see my disclaimers for similar epithets employed as protests against assisted dying, Chapter 11).

But here (in parenthesis), we should not forget that assisted deaths are thought reasonable by those whose definitions of human personhood are exclusively directed towards a selective 'moral community' of actively-engaged people with memories and intentions, as variably promoted by Singer, Lockwood, Tooley and others. Another approach urges legalised assisted dying on grounds that animals, for whom scientific accounts are emphatic of their consciousness and sensitivity to pain (as Green agrees), can be easily 'put down' when thought to be *in extremis*. Yet no (UK) vet is subject to laws prohibiting such acts, and if an animal is killed for no valid reason, posthumous forensic accounts are neither required, nor demanded by law. However, for both groups, there is a need to overcome the conundrum of a legally based and legally enforced duty of care by all professionals working surgically, for example, with human foetuses *in utero*, or with extremely early premature babies in neonatal intensive care units. The law here assumes that these humans, although unable to speak up for themselves, *are persons in their own right* and hence demand appropriate levels of responsible competence, respect, and management.

Another crucial difficulty with 'image' usage involves children who have incurred severe brain damage or are subject to the consequences of a major metabolic abnormality, both invariably revolting for onlookers. In their eyes, such horrors would hardly equate with the Edenic vision of a 'perfect world' arising from our supposedly God-given creation, and 'made' in

God's image. Of course, these gross anomalies do not negate the use of the image theme, as such, but they certainly condition its usage, thus disturbing what is usually taken to imply high-level concepts of humanness and associated manifestations of being specifically, if not uniquely, favoured by the Godhead.

Second, in dealing with the 'image' problem, van Huyssteen (his book, 294) turns to contemporary Jewish writers for inspiration. But modern Jewish scholars, like other contemporary colleagues, will surely be subject to the same restrictions in prising open what the original Genesis writers (and their later redactors) had in mind when using the Hebrew words *demut* and *tselem*. Their meanings as image or likeness, and the subtle differences and nuances, are not so easily discerned, nor why *demut* is subservient to *tselem*. Professor James Barr pointed out that these words were employed despite the stern embargo on image usage, utterly and repeatedly proscribed by Deuteronomy (4:16,13,15). Therefore, in my view, image-of-God themes cannot be usefully employed, critically, as a methodology for deriving fresh meanings to the conundrum of defining persons. Neither do I see how science could possibly explicate its meaning through van Huyssteen's 'traversal' conversation, since science is quite unable to postulate a substantive answer to a biblically-based problem. So his project fails on this crucial issue.

Third, if any clarity is to be had on the meaning of *imago Dei*, then it might be more usefully insisted that Jesus, as perfect Man, alone, fulfils those criteria: 'He is the effulgence of God's glory and the exact representation of his being' (Heb 1.3). As created individuals, we cannot ascend to that lofty role. While Genesis commands us to be the divine stewards of the earth, acting as His vice-regents, the annals of recorded history enunciate mankind's grossest failures in achieving that divine command. Furthermore, man's inhumanity to man continually causes us to shrink away in horror and disgust (Disability: Chapter 8). There can be no doubt about the trail of disasters which litter our history (both to

232

mankind, and to the world and its resources) and which show scant hint of abating. That others think differently is, to my mind, somewhat perverse and inconsistent with man's intrinsic violent nature.[8,9,10] Surely the problem reflects our inbuilt (animal) natures which can never be contained by society's rules and strictures and nor, it appears, by divine prohibitions or decrees either.

These problems represent some of the frustrations of being created, of being formed in flesh which, by definition, cannot equate to divine purity and holiness, or even a reflective 'image' of the Godhead. The starting point, as John Cottingham elucidates, is that of metaphysical evil, or of an original, inbuilt imperfection in the world.[11] God cannot create something which is identical to his perfectness, and therefore the world – not being of himself – must be distinct and discernible, and thus also less than perfect. And because all of us as created living things, are not free, in the absolute sense, we are unable to get outside ourselves or even function from any point-of-view beyond the world. Thus, we can never have an independent view or function from an external 'nowhere', but only what can arise within our habitation on earth.

Furthermore, although this strays away from the main theme, the earth as a *living* planet must, as earth scientists tell us, undergo tsunamis, earthquakes, volcanic eruptions, floods, tornados and ice-ages as part of its normal 'geo-physiology'. Such natural catastrophes inevitably incur considerable losses of innocent life. For those people involved, where no moral blame is relevant, yet who are mercilessly excised from their lives, environment and families, surely some kind of redress is necessarily important.[12] And, if we were to press this line of argument a little further, some respite is surely necessary for those inflicted with evolutionary genetic mutations which rob them of life and the prospect of living it to the full.

Finally, we should not dismiss the endless tales of childhood abuse by priests, religious (educational) orders, and even those in charge of care homes. Nowadays, it is so common to hear stories from grown men about their revulsion at how they were so cruelly and mercilessly abused by those 'in authority' as young-sters. Their lives have been wrecked, and the memory of such assaults hangs heavily with them. For this growing contingent, there are no sweet memories of a carefree, loving childhood projected into adulthood. So I challenge those philosophers who think memory is such an important attribute of personhood to think again and to widen their horizons. These stories are not a few, crazy, manifest aberrations, but apply to a growing number of real people with real, lifelong hurts destructive of their oppor-tunity to gain any happiness resulting from hauntingly past encounters.

The Unknowingness of Our Createdness

In pursuing further another problem of createdness, it becomes apparent that we can never know ourselves to our ultimate depths, and therefore cannot begin to attempt to make sense of our essential 'who-ness', nor, importantly that of others.

'The question most central to recognition is a direct one, and it is addressed to the other: 'Who are you?'', writes Judith Butler: 'This question assumes that there is an other before us whom we do not know and cannot fully comprehend, one whose uniqueness and non-substitutability set limits [on] reciprocal recognition...', and this in common with Pannenberg who also, rightly, makes the same point.[13,14] In other words, if we were able to take people apart, like an engineer disassembling a washing machine or a car engine, they would no longer be sources of wonder, fascination, intrigue, elusiveness – even mystery. Instead we would be regarded solely as objects of disdain, to be scorned at, rejected, passed over, or tossed aside like dolls. That is not only important for us as human beings – but most worrying, because it means we cannot grasp to the fullest extent the essential 'being' of those nearest to us and whom we love the most.

In my experience, that sense of unknowing is manifested when one's grown children visit for a day with spouse and offspring. There is talk, laughter, play, and food, but the conversation deals largely with 'what-ness' – job, house, schools, children, car and so on. But once the time of departure has passed, all that remains is that hollow feeling of remorse and loneliness because they have gone and are now absent – forcing the realisation that nothing more has been learned about *who* these people are; particularly in respect of one's own child.

These are all aspects of what Zizioulas calls the paradox of life's *presence-in-absence*, a state also realised at any travel terminus when the overwhelming thought is of the anticipated appearance of one's beloved, but who fails to arrive on time, as scheduled.[15] We have seen the same paradox in operation over the two recent aircraft losses in the Far East. The faces of relatives are focussed solely on the last recalled memories of those who were passengers: yet the actual people (and indeed their bodies) are absent – and unaccountable.

In a more intimate setting, we undergo the same sense of unknowing, of presence-in-absence, as Roger Scruton's account of sexuality poignantly reminds us.[16] In that act, we appear before the other naked, so to remove all outer pretensions employed in our display to the world. And so in our privacy, the fleshly caresses of desire delineate the other's body, thereby revealing the person as embodied – through a glance – or with the other sensual nuances which 'incarnate' the other, especially through the face, the eyes, even a blush. Yet at its height, I cannot know what has been felt, undergone, what I have inaugurated and brought to completion: neither can each narrate to the other that scintillating intensity which floods the body, undergone in that most sacred moment of person-to-person encounter. In the most intimate bodily connectedness and presence, there is an absence, an opacity towards an understanding of the other. A similar forsakenness accompanies death. At funerals the body *although present*, is a lifeless corpse reposing in a coffin. Mourners' thoughts are focussed on 'celebrating' that life now passed, its completeness, particularity, and former societal contribution, *but – in absence*. It is a loneliness that has accompanied and haunted the mystic's 'long night of unknowing'.

So also Frances Young: 'For the most part, our experience of God is of partial presence or apparent absence'.[17] From all this, our inevitable conclusion is that ultimately we cannot know God, ourselves, or others. Even Jesus' acute sense of God's absence is all too disturbingly real (Mt 27.46): 'Oh God, My God, why have you forsaken me?' We must also realise that other occasions which generate similar feelings of intense loneliness cannot be completely assuaged by those whom we love or are nearest to, because they are in the same mould, from which they cannot escape. These recurrent incidents throughout life generate feelings of sorrow, total isolation, foreboding, even abandonment. From this, all too often, suicidal despair sets in and can obviously be fateful unless recognised and dealt with.

Frustrated Transcendence

Despite our frustrations at being 'caged' within our environment, we try to burst out, and in part, we successfully do that. This attempted movement away from the boundaries imposed by being-in-the-flesh has been termed 'transcendence' by John Macquarrie, 'ec-stasis' by John Zizioulas, or 'eccentric existence' by David Kelsey.[18,19,20] Our interrogations of the known universe provide one notable source of wonder, exemplary of what we have been able to establish and understand. But there are many other examples. Some with terminal diseases begin a most energetic fight-back – towards those in a similar predicament, for their own lives remaining, and their now special relationship and loving feelings towards spouse and children. That was the example set by a previous British prime ministerial aide who, dying from cancer aged 36, had already pre-recorded book readings to be played for her young children at bedtime and purchased all the books they should read as they grew into adolescents, while also ascending to the 'strange, razor-like clarity' which occupied her dwindling days here on earth, and contrary to many who would be unlikely to experience the same heightened awareness during their lifetimes.[21]

Most particularly, however, I am reminded of the British virtuoso violinist, Nigel Kennedy. In my view, Kennedy is a non-self-aggrandising, non-self-inflating virtuoso and genius. In a recent interview with BBC television, he declared his violin to be 'his religion' and that he was 'no longer satisfied' simply to play the written notes in correct order and tuning. His improvisatory impulse to break out from the merely written, composed note illumines his transcendent fire, an emergent passion seemingly uncontainable by the human frame. That is evident from the first sounds emitted as bow engages string, and the enraptured facial expression of transcendent joy evoked by the emerging sounds he creates. In passing, we might wonder if this state of mind has any resemblance to being in heaven.

Such a moment of ecstatic sublimity is reflected elsewhere in a poem by the American spitfire pilot John Gillespie Magee (killed in an air collision, 1941). As he put it, he 'chased the shouting wind along...up, up the long, delirious, burning blue...[and] topped the wind-swept heights' there to tread 'the high untrespassed sanctity of space', and – having put his hand out – 'touched the face of God!'[22]

But yet, that is ultimately insufficient, as the well-known Narnia author and Oxford don Clive Staples Lewis discovered during his lifetime at Magdalen College. His successive returns to the worldly creations of Northern (Scandinavian) folk-literature and Wagnerian music progressively lost their 'stab' of joy.[23] We always have that feeling of never being satisfied and forever yearning for more – indeed, for perfect, untrammelled freedom: but that simply is not possible. The late Jesuit scholar, Fr Gerard W. Hughes SJ, elucidates these urges in terms of desire, in his book *The God of Surprises*.

Our desires are not, as often wrongly thought, always selfish and inward looking as vindications for fulfilling inbuilt narcissistic tendencies, but an almost frantic searching into the depths of our innermost selves, or the world, or others, so to find complete satisfaction. We can never achieve all that we wish to complete, and we understand the futility of being cut down with our life's project knowingly unfinished and incomplete. That encapsulates our incessant searching for freedom and for love, yet if and when supposedly found, cannot offer us ultimate rest and satisfaction. But, as noted in Chapter 6, the only way we can achieve absolute, non-contingent freedom (like Zizioulas' reference to Dostoevsky's Kirilov, in *The Possessed*) is to kill ourselves.[25]

Jan-Olav Henrikson takes this further, for he declares:

Desire (for love) and desire emerging from love disclose me, make me appear as being more than the self-sufficient self. Desire is thus – for ethical, phenomenological, and positive theological reasons – that which disrupts a self-sufficient subjectivity without leaving it behind; that which transcends what is conceived in the rational subject's already given understanding of the world. It reminds us of how our self has its origin in the Other who gives content to this world; in that which creates fissures and disturbances in our complacencies by stirring our desires.[26]

However tragic some of this may seem, nonetheless, there may be another route available and far less destructive. Thus Pannenberg: we need Another to ultimately resolve creaturehood. That ever-present, non-fulfillment 'presupposes something outside [humankind] that is beyond every experience of the world'.[27]

Emerging Beyond Mortal Transcendence

In order to resolve all the varied shortcomings visited above, another personhood (Lat: *persona*, Gk: *hupōstasis*) is only possible, within one's earthly birth-death cycle, through communion with the uncreated, immortal Godhead, thus to fulfil those innermost desire(s) for unachievable love and freedom. That is secured through once-for-all baptism as a gateway to a partially gained, longed-for freedom. That is the ultimate hope of resurrection at the 'end time' (or Eschaton) thereby achieving complete rest (stasis) from all our created frustrations in becoming immortalised – 'for this perishable body must put on the imperishable' (I Cor 15:52). Even then, we cannot be entirely sure what those words actually mean, and hence, how we shall actually appear in the final resurrection.

My feeling is that there is too much anthropomorphism creeping in, possibly incurred by the implications of the Pauline spiritual body (*sôma pneumatikon*), especially when we become aware of the varied properties which are detailed to be in place, such as continuity, memory, sociality, further development, and identity. That is evident from chapters in one of various other texts, dealing specifically with theological and scientific approaches to human resurrection, edited by Peters, Russell & Welker.[28] We must also concern ourselves with the post-crucifixion appearances of Jesus. These latter appearances suggest properties that are seemingly robustly physical, and yet also non-physical. It seems clear that it was necessary for Jesus' body to be seen in order to show that he had survived death. Yet, Jesus' earthly appearances, post-crucifixion, may not be a suitable paradigm, in each of their aspects, upon which to model our own progress after death and into resurrection.

Moreover, no one, to my knowledge, has answered Sarah Coakley's pertinent question as to what happened to Jesus' body after he ascended.[29] It raises the additional question whether a 'body' (of the sort we are accustomed to) is necessary in the resurrection: I know of no sensible reason why this should be. Without being too offensive, I think that most people have overly narcissistic attachments to their physical bodies (perhaps a result of today's emphasis on bodily looks and proportions). Not having a body, for most humans, would impact them as functionally, not to mention emotionally, impossible. Yet Professor Stephen Hawking performs adequately, even as a human, only with a cognitively operative brain surviving out of the bits which don't work, a few eye-blinks, and his world famous voice synthesiser. In summary, we just do not know – it is beyond our surmising.

Above, I referred to Nigel Kennedy and the sublimity of his playing. I am specially thinking of his first recording of the Elgar Violin Concerto, and its second movement, during which he draws the listener ever upwards into the heights of rapture. During those intense moments of ecstatic transcendence, one no longer needs memory, future intentions, even the love of near ones (which the Oxford anthropologist Robin Dunbar puts at less than five people for the most intimate social grouping), while all sense of time vanishes.[30] Could this, then, be a possible foretaste of true immortality, given that being within the actual presence of, and enfolded by, the divine Love of the Godhead would be even more a sublime and all-embracing experience, despite its exceeding present human understanding? I make this supposition because of another acute problem arising from the content of that book already mentioned.

This concerns the notable absence in it of any reference to those who, because of brain trauma or metabolic error, would never have any memories, any sufficiently known or remembered social contacts with others (however loving and caring), no future intentions, and least of all, no sense of personal identity, or, of continuity.[28] Most of those contributors I suspect would never have spent a great deal of time in a hospital, or ever attended many ward rounds – yet this complete agnosia about relevant biomedical knowledge and its important conditioning of our envisioned status in the afterlife, seems to be a glaring omission. So I ask: would we actually, as no longer corporeal entities beyond death, require any more than I have suggested?

Much writing about the future is biased, and tied to a carbon-based existence. I know of no one who has been able to envisage a state of existence not informed by earthly life and the need for a body, and certain 'basic' psychological criteria so to render that existence seemingly meaningful, albeit construed in worldly terms. Moreover, the continued seepage of neurophysiology into theological discourse is unlikely to help, as the brain (or a computer) is carbon-based. The question for these neo-neurotheists is how, perhaps, to conceive of a bodiless state in which the rapturous sight of the 'Beatific Vision' could be enjoyed without accompanying pain and anguish – or even feet on the ground. The vision of the prophet Isaiah (Isa 11:6-9) is a carbon-based pastoral: and if all of creation is to be redeemed, how could a lion be doctored so as not to enrol an adjacent lamb, or calf, into its upcoming lunch schedule? It would no longer be a lion: and would an Ebola virus still be as deadly infectious for humans, in the resurrection? Was Paul right in thinking that *all creation* was groaning for redemption (Rom 8:18-20)? But would that be possible? And if so, how? – given our insights into the millions of species since known to have become extinct, resulting from evolutionary pressure and constraints. I am encouraged by the viewpoint of Philip Rolnick: 'The person who can receive...grace...is more valuable than the whole universe, because the universe is a *what*, not a *who*, and the what exists for the sake of the who', (his emphases).[31] Thinking outside the carbon-based box is, for us, well-nigh impossible.

I think there is an overemphasis on anthropological projection into the next world which may be entirely unwarranted. And this is not will-of-the-wisp daydreaming. For example, according to The Health Research Council of New

Zealand there are 60 million people suffering traumatic brain injuries throughout the world and which, as the most common severe disability of humankind, has important bearings regarding its long-term repercussions on neurological and mental health, especially for memory, ability to act independently, and identity.[32,33] That is one problem that cannot be dismissed.

And in Europe (according to *Altzheimer Europe*), there are up to eight million people with dementia, and in the USA, up to six million sufferers (steadily rising as longevity increases).[34] Those with this particular kind of brain failure die without memory, forward intentions, knowledge of those nearest to them including children, and scant idea of identity, or time. From these brief figures alone, it becomes very clear that the criteria established by humans (however expert in their particular fields) and presumably required of the Godhead to satisfy on our behalf after we have died, are exclusive, somewhat misplaced and possibly irrelevant. Although the Godhead might be able to reconstruct, out of this heavily imposed workload, the previous lives of those with dementia, that would not be at all possible for all those who never had a life because of brain anoxia incurred during pregnancy or the birthing process. Has that difficulty never entered into the thinking of philosophers and their like?

And, in passing, might we also enter a plea for all the two million foetuses aborted across Europe, UK, US, and Canada on an annual basis – and who never quite make it into life and its possibilities?

In the same volume, Nancey Murphy is hence quite correct, given the concerns I have raised, by her warning about over-enthusiastic speculative claims descriptive of the afterlife.[35] And I mean by this – 'who' we shall be, 'what' we shall be like, and 'where' we will be located. She quotes Wittgenstein's seventh proposition that if we don't have the data, or simply don't know, then we would be better to keep quiet.[36]

Following Zizioulas and Pannenberg, I have already suggested that in baptism (Chapter 6: Theological Anthropology) we are partially taken up into the Godhead where we become 'known' and 'sacralised'. Within the nascent 1st-Century church, there was no consistently understood notion of baptism, as Lars Hartman in his detailed analysis, points out.[37] Despite Jesus' death and the shedding of his blood as the one act of atonement 'for all men at all times', the ritual of baptism was taken as a personal redemptive washing from sin. Nonetheless, the baptised subject as everyone realised would clearly sin again, but since baptism was very firmly regarded as a once-in-a-lifetime event, the occasion clearly implied far more, importantly embodying entry into the sacral community of the church, and thus a cleansing from the filth and pollution seen as a contaminant of life outside that community. It thus connoted a cosmic dimension in causing a regeneration (*annagenēsis*) of the person as a son of the Father, as was Jesus (Mk. 1:10-11).

Especially strange, within this new verbal context, were the formulaic 'into Christ', 'in Christ' or even 'into the name of Christ'. As Hartman puts it, the Greek was unbiblical, and would have descended strangely on the ears of the Greek-speaking inhabitants of the Diaspora. These phrases carried with them an echo of death and resurrection, such that the candidate 'died' and in discarding his existing bodily 'corpse', thence became clothed in, or with, Christ. It is quite evident that generally, baptism signified a profound eschatological alterity in rendering the postulant pure and holy, and henceforward now radically changed as a member of the sacred (ecclesial and eucharistic) community on earth, as in heaven. Those central cosmic and eschatological roles of baptism may have been partly lost, in recent times.

In and through baptism we become members of the Eucharistic communion within the earthly church. In baptism we, as Christians, are 'born again', not 'from blood, nor of the will of the flesh, nor of the will of man' (Jn 1:13), but 'from above' thereby becoming 'adopted' as 'children' of God, who is 'Our Father in Heaven', and in whom there 'is perfect freedom'.[38] Additional to base corporality and the inevitable threat of death, we acquire a new personhood, or 'hypostatic relationality' as it is termed theologically, within the Godhead (I Pet 1.23 – *Anagegen-nēménoi*) known as 'theosis', or being 'godlike', within the Orthodox tradition. Thereby, authentic who-ness is partially captured during life, although only ultimately realised as absolute uniqueness beyond death. These often-used scriptural and liturgical aspirations are explicitly about a state of being, an otherworldly state (ontologically and metaphysically): they make little sense, although rarely acknowledged in either the theological literature or from the pulpit, in the physical environment of creatureliness (Chapter 6). Indeed, the important, eschatological role of baptism (in addition to its accompanying historical, cosmic, and personal dimensions as so profoundly expressed by John Zizioulas), hardly gets a mention in so many texts dealing with personhood, while the continued grasp on the soul as agent of immortality cannot, seemingly, be vouchsafed (perhaps because of its last-lingering narcissistic attachments to our perceptions of ourselves as bodies progressing into immortality).[39]

Through baptism and all its inferences about a new type of spiritual personhood adjacent, as it were, to our current physical body and personhood, is an important tenet of the sacramental event. This has to be stressed as well as understood, because through that understanding, we overcome the dilemma of those for whom a physical life has *no* meaning, *no* sense of continuity, and in which there can be *no* hope of any accomplishment whatsoever, based on expectations of intellectual development alone. Baptism into Christ, through the power of the Holy Spirit, raises these deprived and impoverished persons to a state whereby they now have a 'heavenly' placement, as the predominant means (perhaps) of ensuring for them ultimate rest among the Saints in the everlasting (eschatological) Kingdom. We thus acquire two natures within the one person. And the guarantee is the Chalcedonian proposal, which unites the human with the divine in the incarnate Jesus: one person but of two natures, undivided.

Becoming the Real Person

In constituting a very lofty envisioning of the real meaning of personhood, these observations illuminate the uncertainties and ambiguities of earthly life, of being human, yet simultaneously point to a becoming 'beyond the world' in our being 'divinised' within the Godhead (or 'en-hypostatised'): in other words, the ecstatic glimpse of the infinite, and thus of God. And as such, this resolves the final paradoxes of life, of death, and of the accompanying fear of death. The Orthodox concept of being divinised or achieving 'theosis' overcomes Platonic notions of a soul floating out of a body, and the ultimately unresolved meaning – even an intelligent understanding – of being made in God's image, other than being drawn towards, and into, the Godhead. This is part of the uncertainty that remains over van Huyssteen's presumptions. Even on the last page of his book, he is still querying how the scope of an inter-disciplinary dialogue between science and theology could resolve the problem of human uniqueness and what it means: I hardly think that could be accomplished from within the created world and based on worldly characteristics.[40]

At the beginning of this book, I suggested that thinking about mankind, in terms of the made-in-God's-image theme, was not a promising venture. Indeed, I have often wondered whether, if the first two chapters of the Genesis scroll had been lost and never read, our perceptions of mankind's status within the created world would have been significantly changed. The grandeur implied by the made-in-God's-image easily leads to an over-inflated perception of our roles in the world. Moreover, given the recurrent atrocities that litter humankind's history and our relentless exploitation of all the reserves which the world offers, we have not fulfilled that divine calling with honour. It is also significant how, throughout the Hebrew scriptures, scant reference is made to this particular form of calling by the Godhead.

On the other hand, another model can be based on baptism. John's baptism with water for repentance (a metanoia, or reorienting of ourselves towards the Godhead) represented a significant turning away from the Jewish bath (or mitzvah) for body cleansing purposes only. Nevertheless, John acknowledged that his baptismal approach was incomplete and would soon be superseded by that of Jesus, in being accompanied by the additional fire of the Spirit. The use of oil and water indicate the dual action involved in the baptismal ceremony.

Through baptism we now become known to the Godhead, which guarantees for us a partial foothold in the Kingdom – beyond the spectre of dying. Our belief and hope is in the death and resurrection of Jesus who, in proleptically demonstrating that death no longer held dominion over him, thereby demonstrates and establishes identical criteria for us. By his resurrection appearances, and ultimate ascension, he revealed beyond any doubt that he had taken the sting out of death and dying, not only in making it actually intelligible – but even in rendering the prospect bearable.

This seems to provide a far better nuanced explanation for the observation that, despite our highly evolved cognitive prowess, we do not know ourselves to our very depths, nor those with whom we have intimate relationships. Indeed, cognitive competence is barely sufficient to frame humankind, despite van Huysteen's judgement. True, baptism, as well as eucharistic practice, bring us into communion with the Godhead. Despite its partial role, while we still remain attached to our bodies, baptism confers on mankind a status which, by making us 'en-hypostatised' within the Godhead, renders us 'worthy of our calling' despite all our abject failings.

Thus I am in agreement with the insights provided by Catherine Keller, in reminding us of the proper translation of the Pauline text (1 Cor 13:12).[41] She recalls the Greek and its subtleties, based on Conzelmann's exegesis of First Corinthians: 'Now we see in a mirror, in an enigma (enigmatically, indistinctly, or with uncertainty), but then, person to person'.[42]

First, she reminds us that the word is a mirror (*ěsoptron*), but not the familiar smooth silvered surfaces known today: nor can it simply be conceived of as just a 'glass' – even smoked. It is more like the rippling of a soft breeze over water in which we perceive ourselves – but enigmatically, wondering who this image actually reflects; a rather obscure riddle, fractured, uncertain, puzzling.

Second, the usual 'face' is replaced by a proper rendering of *prosopon* – person. In classical Greek theatre, the prosopon was a mask employed in shielding the true identity of the actor who thus could say what he liked without fear of repercussion. And so, in looking into the mirror, we are confronted by that enigmatic mask, covering up who we really are, or might be, and thus camouflaging the public make-up of how we wish to present ourselves: but we can't be sure because we really don't know.

Third, in that face-to-face, or person-to-person encounter, Keller suggests we may not even have found the certainty of God (her chapter, p. 301-2). But Paul probably had in mind the Hebrew (similar to the modern Israeli *liphnē*), whereby the grammatical Semitic idiom 'before the face of' (God, or whoever) actually means standing directly in front of that individual, so that face-to-face would be as good a translation as person-to-person. We know what it is like to be face-to-face with someone else. Observed in that sense, as Judith Butler recognises, we can be undone by the other in any relational stance.

That sums up our earthly context, and one which science can neither usurp nor even attempt to offer further explanatory insights into. We cannot know God ourselves, or those with whom we are in the closest relationships. That is both the tragedy and the paradox of being here, and the only partial answer, as S. Paul urges, is to be in the Godhead, to have died and risen with Christ, to walk in the ways of the Kingdom, and to attempt to follow its ethical demands. In other words, to undergo a determined *metanoia* or a turnabout in our lives, in order to be aligned with that ethos of deep care and respect for the other, as demanded by Kingdom ethics.

To have been baptised is the means known to us for being recognised as special before our adopted God-as-Father, in realising our true 'ipseity', our unique 'incommunicabilis'.[43] It renders all of us considerable, and therefore valuable, cherished and holy in his sight.[44] Then we become someone that actually is *unique* and *unsubstitutable*, then we come face-to-face with, and in, and before God-as-Father: then, finally, we come to know our true selves – precious and unrepeatable. And, in having been redeemed through the Son's sacrifice upon the cross, we are thereby guaranteed our rightful place in heaven by his resurrection from the dead, and through death we, ourselves, will also put on immortality.

The symbolism of baptism promises that the temporary state of a transformed 'person', incomplete as it is in life as associated with its contingent biological or fleshly being will become – through a newly creative act of God – the basis of the new resurrection entity with, and in, Christ and the Spirit.[45] How we shall be changed is unknown, and neither do we have much of a clue as to what type of beings we shall be in the New Creation, nor to what extent we shall be 'embodied' – other than within the Godhead.

While the baptismal act guarantees a genuine continuity with our present existence, ultimate resurrection in the end days of the

Eschaton will entail the loss of many of our worldly ties – mental, physical, and relational, once we have died and taken up our abode in the New Creation when assuming our unique resurrectional 'self'. There will be an abrupt discontinuity with our previous worldly life; no smouldering embers or a lingering afterglow, and no 'soul' fluttering away like a final-gasp butterfly appearing out of a corpse-like shrivelled chrysalis.[46] And as Paul Fiddes continues: if resurrection is centred in, and with, and on Christ 'then there will be no separate resurrection bodies existing on their own: we shall be inseparable from the inclusive resurrection of Christ, which has cosmic scope'.[47]

Only in steadfast faith borne up by our increasing transcendent outreach to the divine hope and promise, granted to us in prophecy, true revelation and grace sacramentally received, can we be assured that we shall eventually arrive at our destination. This may well require having to ditch much of our long-accumulated, cherished earthbound idiosyncrasy, props and luggage – a prospect that while we are alive may seem an unpleasant necessity – if not somewhat threatening. The least we might expect, however, is that we shall come face-to-face with the reality of the triune Unity of our former desires, hope and faithfulness, as anticipated most simply, yet in quiet confidence, by the Psalmist:

For myself, in righteousness, shall I gaze upon your face: then, in my awakening, shall I come to be replete within your image (Ps. 17.15: my translation).

* * *

1. Huxley A, 1928. *Point-Counter-Point*. London: Penguin.
2. Haeckel E, 1866. *Generalle Morphologie der Organism*. Vol 2. Berlin: Reimer.
3. Tillich P, 1997. *Systematic Theology* (Vol I). London: SCM

Xprint 164.

4. van Huyssteen JW, 2006. *Alone in the world? Human Uniqueness in Science and Theology.* Grand Rapids: Eerdmans.

5. Green J, 2008. *Body, Soul, and Human Life.* Grand Rapids: Baker Academic.

6. Marsh MN, 2012. *The Moral Status of the Embryo-Foetus: Biomedical Perspectives.* (Ethics #166). Cambridge: Grove Publications 15-17.

7. Barr J, 1968/9. The image of God in the book of Genesis – a study of terminology. *Bull John Rylands University Library (Manchester)* 51: 11-26.

8. Rifkin J, 2009. *The Empathic Civilisation: The Race towards a Global Consciousness.* Cambridge: Polity Press.

9. Pinker S, 2011. *The Better Angels of our Nature: The Decline of Violence in History and its Causes.* London: Allen Lane.

10. Muchembled R, 2011. *A History of Violence: From the End of the Middle Ages to the Present.* Cambridge: Polity Press.

11. Cottingham J, 2005. *The Spiritual Dimension.* Cambridge: Cambridge University Press 27ff.

12. Adams MM, 1999. *Horrendous Evils And The Goodness Of God.* London: Cornell University Press.

13. Butler J, 2005. *Giving an Account of Oneself.* New York: Fordham University Press 31.

14. Pannenberg W, 1972. *What is Man?* Philadelphia: Fortress Press.

15. Scruton R, 1996. Sex. In: *An Intelligent Person's Guide to Philosophy.* London: Duckworth 128-129.

16. Zizioulas J, 1975. Human capacity and incapacity: a theological exploration of personhood. *Scot J Theology* 28: 401-447.

17. Young F, 1984. *Can These Dry Bones Live?* London: SCM Press 57-59.

18. Macquarrie J, 1983. *In Search of Humanity: A Theological & Philosophical Approach.* New York: Crossroad 25-37.

19. Zizioulas, incapacity 408-410.
20. Kelsey D, 2009. *Eccentric Existence: A Theological Anthropology.* Louisville: John Knox Press 275; 284ff; 1045-1051.
21. Gross K, 2014. *Late Fragments: Everything I Want To Tell You (About This Magnificent Life).* London: Collins 145; 227-8.
22. Magee JG & Garnett S, 1989. High Flight. In: *The Complete Works Of John Magee, Pilot Poet.* Cheltenham (UK): This England Books.
23. Lewis CS, 1977. *Surprised by Joy.* London: Fount Books.
24. Hughes GW, SJ, 1985. *The God Of Surprises.* London: Darton, Longman & Todd 94-100.
25. Henrikson J-O, 2011. The Erotic Self And The Image Of God. In: Huyssteen W & Wiebe E (eds), *In Search Of Self: Interdisciplinary Perspectives On Personhood.* Grand Rapids: Eerdmans 261.
26. Zizioulas J, 2011. *The Eucharistic Communion And The World.* London: Clark 168.
27. Pannenberg, *Man?* 10-11.
28. Peters T, Russell RJ, Welker M, 2002. *Resurrection: Theological And Scientific Assessments.* Grand Rapids: Eerdmanns.
29. Coakley S, 2002. What does Chalcedon solve and what does it not? Some reflections on the status and meaning of the Chalcedonian 'Definition'. In: Davis S, Kendall D, O'Collins G (eds), *The Incarnation.* Oxford: Oxford University Press 162-163.
30. Dunbar R, 2014. *Human Evolution.* London: Pelican 79-80.
31. Rolnick P, 2007. Gift: Summoned, Interrogated, Enjoyed. In: *Person, Grace, and God.* Grand Rapids: Eerdmanns 169.
32. Feigin V, Theadom A, Barker-Collo S, et al, 2013. Incidence of traumatic brain injury in New Zealand: a population-based study. *Lancet Neurology* 12: 53-64.
33. Lathif N, Phipps E, Alton P, Sharma D, 2014. Prevalence of psychiatric disorders following brain injury. *Br J Med Pract* 7: a722.

34. Alzheimer Association (USA), 2014. *2014 Alzheimer Disease Facts & Figures*. Alzheimer's & Dementia 10.

35. Murphy N, 2002. The Resurrection Body And Personal Identity: Possibilities And Limits Of Eschatological Knowledge. In: Peters T, Russell RJ, Welker M, *Resurrection* 202.

36. Wittgenstein L, 1961. *Tractatus Logico-Philosophicus*. USA (Pbk) 96.

37. Hartman L, 1997. *Into The Name Of The Lord Jesus*. Edinburgh: Clark.

38. Zizioulas J, 2006. On Being Other: Towards An Ontology Of Otherness. In:McPartlan P (ed), *Communion And Otherness: Further Studies In Personhood And The Church*. London: Clark 13-98.

39. Zizioulas, *Eucharistic Communion*, 116-8.

40. van Huyssteen, *Alone* 325.

41. Keller C, 2011. Enigmatic Experiences: Spirit, Complexity, and Person. In: van Huyssteen W & Wiebe E (eds), *In Search Of Self: Interdisciplinary Perspectives On Personhood*. Grand Rapids: Eerdmans 301-318.

42. Conzelmann H, 1975. *I Corinthians: A Commentary on the First Epistle to the Corinthians* (ET: JW Leitch). Philadelphia: Fortress Press 226-229.

Conzelmann points out that the mirror analogy and the idea of confrontation probably has nothing to do with Corinth being the place of mirror production, but reveals an extensive history related to ancient sources throughout literature, secular and biblical (Num 12:8; Gen 32:30).

43. Rolnick, *Person*, 40ff.

44. Goodpaster K, 1978. On being morally considerable. *J Philosophy* 75: 308-25.

45. Zizioulas J, 2006. *Communion & Otherness* (ed. McCartlan P). London: Clark 79-81;188;277-283.

46. Muddiman J, 1994. 'I believe in the Resurrection of the Body'.

In: Barton S & Stanton G (eds), *Resurrection*. London: SPCK 135.

47. Fiddes P, 2000. *The Promised End: Eschatology In Theology And Literature*. Oxford: Blackwell 53; 76; 98.

Bibliography

Adams MM, 1999. *Horrendous Evils And The Goodness Of God*. London: Cornell University Press.

Adams MM, 2006. *Christ And Horrors: The Coherence Of Christology*. Cambridge: Cambridge University Press.

Adams MM & Adams RM (eds), 1990. *The Problem of Evil*. Oxford: Oxford University Press.

Allen GE, 2004. Was Nazi eugenics created in the US? *EMBO Reports* 5: 451-2.

Alzheimer Association (USA), 2014. *2014 Alzheimer Disease Facts & Figures*. Alzheimer's & Dementia 10.

Anderson S, Bechara A, Damasio H, Tranel D, Damasio A, 1999. Impairment of social and moral behaviour related to early damage in human prefrontal cortex. *Nat Neurosci* 2: 1032-1037.

Anton S, Potts R, Aiello L, 2014. Evolution of early *Homo*: an integrated biological perspective. *Science* 344: 1236828-1236828-13.

Arendt H, 1959. *The Human Condition*. New York: Double Day.

Ashby G, 1988. *Sacrifice: Its Nature and Purpose*. London: SCM Press.

Assisted Dying Bill [HL Bill 24]: London: The Stationery Office, 15 May 2013.

Ayub Q, Yngvadottir B, Chen Y, et al, 2013. FOXP2 targets show evidence of positive selection in European populations. *Am J Hum Genet* 92: 696-706.

Badham P, 2009. *Is there a Christian Case for Assisted Dying?* London: SPCK.

Badham P, 2015. Assisted Dying: For and Against. *Modern Believing*, Vol. 56 (2).

Barnes C, 2008. Generating change: disability, culture and art. *Behinderung und Dritte Welt* 19: 4-13.

Barr J, 1968/9. The image of God in the book of Genesis – a study of terminology. *Bull John Rylands University Library (Manchester)* 51: 11-26.

Baughen M, 2010. *The One Big Question.* Crusade World Revival.

Beck A, Kovacs M & Weissman A, 1975. Hopelessness and Suicidal Behaviour: An Overview. *J Am Med Assoc* 234: 1146-9.

Bennett R, 2011. 'Toddler's killers were known as a danger to children'. *The Times* (London), 2011.

Berger L, de Ruiter D, Churchill S et al, 2010. *Australopicethus sediba*: a new species of Homo-like Australopith from South Africa. *Science* 328: 195-204.

Berger P, 1969. *A Rumour of Angels.* London: Penguin.

Berwick R, Friederici A, Chomsky N, Bolhuis J, 2013. Evolution, brain, and the nature of language. *Trends Cogn Sci* 17: 89-98.

Biederman I, Kim J, 2008. 17,000 years of depicting the junction of two smooth shapes. *perception* 37: 161-164.

Biggar N, 2004. *Aiming to Kill: The Ethics of Suicide and Euthanasia.* Cleveland: Pilgrim Press.

Bok S, 1993. Impaired physicians: What patients should know. *Cambridge Quarterly of Healthcare Ethics* 2: 331-340.

Bok S, 1998. There is every reason to look with wary eyes at any calls to put much faith in the power of 'appropriate legal safeguards'. In: Dworkin G, Frey RG, and Bok S (eds), *Euthanasia and Physician-Assisted Suicide.* Cambridge: Cambridge University Press.

Boly M, Phillips C, Tshibanda L, et al, 2008. Intrinsic brain activity in altered states of consciousness: How conscious is the default mode of the brain? *Ann N Y Acad Science* 1129: 19-129.

Boycott B, Young J, 1955. A memory system in Octopus vulgaris Lamarck. *Proc R Soc London B* 143: 449-480:

Braun FM, 1960. *Jean Le Théologien.* Paris: Gabalda.

Brown D & Loades A (eds), *Christ: The Sacramental Word.* London: SPCK.

Brugger P, Kollias S, Muri R, et al, 2000. Beyond remembering:

phantom sensations of congenitally-absent limbs. *Proc Natl Acad Sciences [USA]* 97: 6167-6172.

Buber M, 1959. *I and Thou* (2nd Ed). Edinburgh: Clark.

Buckner R, Sepulcre J, Talukdar T, et al, 2009. Cortical hubs revealed by intrinsic functional connectivity: mapping, assessment of stability, and relation to Alzheimer's disease. *J Neuroscience* 29: 1860-1873.

Butler J, 2005. *Giving an Account of Oneself.* New York: Fordham University Press.

Cabanac M, Cabanac AJ, Parent A, 2009. The emergence of consciousness in phylogeny. *Behav Brain Res* 198: 267-272.

Calcagno JM, Fuentes A, 2012. What makes us human? Answers from evolutionary anthropology. *Evol Anthropol* 21: 183-194.

Cavalieri P, Singer P (eds), 1993. *The Great Ape Project.* New York: St. Martin's Press.

Cerling T, Manthi F, Mbua E, et al, 2013. Stable isotope-based diet reconstructions of Turkana Basin hominins. *Proc Nat Acad Sciences [USA]* 110: 10501-10506.

Chalmers D, 1997. *The Conscious Mind.* Oxford: Oxford University Press 1997.

Chen S, Krinsky B, Long M, 2013. New genes as drivers of phenotypic evolution. *Nature Rev Genet* 14: 645-660.

Cheyne J, Meschino L, Smilek D, 2009. Caricature and contrast in the Upper Palaeolithic: morphometric evidence from cave art. *Perception* 38: 100-108.

Chochinov H, Wilson K & Enns M, 1998. Depression. Hopelessness, and Suicidal Ideation in the Terminally Ill. *Psychosomatics* 39: 366-70.

Choi C, 2013. Early human diets. *Proc Nat Acad Sciences [USA]* 110.

Clarke D, Kissane D, 2002. Demoralisation: its phenomenology and importance. *Austr N Z J Psychiatr* 36: 733-42.

Clément O, 2000. *On Human Being – A Spiritual Anthology* [ET]. New York: New City.

Clough AH, 1862. 'The Last Decalogue'. In: Mulhauser FL, 1974. *The Poems of Arthur Hugh Clough* (2nd Ed). Oxford: Oxford University Press.

Coakley S, 2002. What does Chalcedon solve and what does it not? Some reflections on the status and meaning of the Chalcedonian 'Definition'. In: Davis S, Kendall D, O'Collins G (eds), *The Incarnation*. Oxford: Oxford University Press

Conzelmann H, 1975. *I Corinthians: A Commentary on the First Epistle to the Corinthians* (ET: JW Leitch). Philadelphia: Fortress Press.

Cottingham J, 2005. *The Spiritual Dimension: Religion, Philosophy and Human Value*. Cambridge: Cambridge University Press.

Cox JA, 2011. Disability as an Enacted Parable. *J Religion Dis & Health*. 15: 241-253.

Crick F, 1994. *The Astonishing Hypothesis*. London: Simon & Schuster.

Cullman O, 1967. Immortality of the Soul or Resurrection of the Body. In: Stendahl K (ed), *Immortality & Resurrection*. New York: Macmillan.

Curtiss S, Fromkin V, Krashen S, et al, 1974. The linguistic development of Genie. *Language* 50: 528-554.

Dahl, ME, 1962. *The Resurrection of the Body*. London: SCM.

Damosieaux J, Rombouts S, Barkof F, et al, 2006. Consistent resting-state networks across healthy subjects. *Proc Nat Acad Sciences [USA]* 103: 13848-13853.

Darley JM, Batson CS, 1973. 'From Jerusalem to Jericho': a study of situational and dispositional variables in helping behavior. *J Pers Soc Psychol* 2: S100.

de Castro J, Rosas A, Carbonell E, et al, 1999. A modern human pattern of dental development in lower Pleistocene hominids from Atapuerca-TD6 (Spain). *Proc Nat Acad Sciences [USA]* 96: 4210-4213.

de Leon M, Golovanova L, Doronichev V, et al, 2008. Neanderthal brain size at birth provides insights into the evolution of human life history. *Proc Nat Acad Sciences [USA]* 105: 13764-13768.

de Menocal P, 2014. Where we came from. *Scientific American* 311: 33-37.

De Silva L, 1979. *The Problem of the Self in Buddhism and Christianity.* London: Macmillan.

DeCasper A, Fifer W, 1980. Of human bonding: newborns prefer their mothers' voices. *Science* 208: 1174-1176.

Dediu D, Levinson S, 2013. On the antiquity of language: the reinterpretation of Neanderthal linguistic capacities and its consequences. *Front Psychol* 4: 1-17.

Deech R & Smajdor A, 2010. *From IVF to Immortality.* Oxford: Oxford University Press.

Douglas M, 1966. *Purity & Danger.* London: Routledge.

Dunbar R, 2014. *Human Evolution.* London: Pelican.

Dunn JDG, 1998. *The Theology of Paul the Apostle.* Edinburgh: Clark.

EHRC, London: 2009. *Promoting the Safety and Security of Disabled People.*

EHRC, London: 2011. *Hidden in Plain Sight: Inquiry into Disability-Related Harassment.*

Elliott C, 1992. Diagnosing blame: responsibility and the psychopath. *J Med Philosoph* 17: 200-214.

Feigin V, Theadom A, Barker-Collo S, et al, 2013. Incidence of traumatic brain injury in New Zealand: a population-based study. *Lancet Neurology* 12: 53-64.

Fiddes P, 2000. *The Promised End: Eschatology In Theology And Literature.* Oxford: Blackwell.

Figueiredo J, Frank J, 1982. Subjective incompetence, the clinical hallmark of demoralisation. *Compr Psychiatr* 23: 353-63.

Finlay I, 2009. Hansard HL Deb, Vol. 712, Column 607, 7th July.

Fisher S, Marcus G, 2005. The eloquent ape: genes, brains and the evolution of language. *Nature Rev Genet* 7: 9-20.

Fisher S, Scharff C, 2009. FOXP2 as a molecular window into

speech and language. *Trends Genet* 25: 166-177.

Fitch W, 2000. The evolution of speech: a comparative review. *Trends Cogn Sci* 4: 258-267.

Ford N, 1988. *When did I begin?* Cambridge: Cambridge University Press.

Foster J, 1985. Personhood and the Ethics of Abortion. In: Channer JH (ed), *Abortion & the Sanctity of Human Life.* Exeter: The Paternoster Press.

Fradd CR, 1985. An Introduction to the History and Present State of the Law relating to Abortion in England. In: Channer JH (ed), *Abortion & the Sanctity of Human Life.* Exeter: The Paternoster Press.

Frankl V, 1985. *Man's Search for Meaning.* New York: Pocket Books.

Fransson P, Skiold B, Horsch S et al, 2007. Resting-state networks in the infant brain. *Proceedings National Academy Sciences USA* 104: 15531-15536.

Frasson P, Marrelec G, 2008. The precuneus/posterior cingulate cortex plays a pivotal role in the default mode network: evidence from a partial correlation network. *Neuroimage* 42: 1178-1184.

Frith C, Frith U, 1999. Interacting minds – a biological basis. *Science* 266: 1692-1695.

Galton F, 1904. Eugenics: its definition, scope, and aims. *American J Sociology* 10: 1-25.

Giacino J, Ashwal S, Childs N, et al, 2002. The minimally conscious state: definition and diagnostic criteria. *Neurology* 58: 349-353.

Gibbons K, 2014. How Robert Peston coped with his 'Overwhelming Loss'. London: *The Times*, 1st February 36.

Gibson J, 1998. *Language and Imagery in the Old Testament.* London: SPCK.

Gollwitzer H, Kuhn K, Schneider R, 1974. *Dying We Live.* London: Fontana (German text).

Goodpaster K, 1978. On being morally considerable. *J Philosophy* 75: 308-25.

Gordijn B, 1999. The troublesome concept of the person. *Theoret Med Bioethics* 20: 347-359.

Gowlett J, Gamble C, Dunbar R, 2014. Human evolution and the archaeology of the social brain. *Current Anthropology* 53: 693-722.

Green J, 2008. *Body, Soul, and Human Life*. Grand Rapids: Baker Academic.

Green R, 2002. Part III – Determining Moral Status. *Am J Bioethics* 2: 20-30 (22).

Green R, Krause J, Briggs A, et al, 2010. A draft sequence of the Neanderthal genome. *Science* 328: 710-722.

Greenfield S, 1998. *The Brain – A Guided Tour*. London: Phoenix.

Griffin DR & Speck GB, 2004. New evidence of animal cognition. *Anim Cogn* 7: 5-18.

Gross K, 2014. *Late Fragments: Everything I Want To Tell You (About This Magnificent Life)*. London: Collins.

Gunton C, 2002. *The Christian Faith*. Oxford: Blackwell.

Gunton CE, 1991. Trinity, Ontology and Anthropology: Towards a Renewal of the Doctrine of the *Imago Dei*. In: Schwöbel C & Gunton CE (eds), *Persons Divine and Human*. Edinburgh: Clark.

Gusnard D, Raichle M, 2001. Searching for a baseline: functional imaging and the resting human brain. *Nat Rev Neuroscience* 2: 685-694.

Haas P, 1992. *Morality after Auschwitz: The Radical Challenge of the Nazi Ethic*. Philadelphia: Fortress.

Haeckel E, 1866. *Generalle Morphologie der Organism*. Vol 2. Berlin: Reimer.

Hampson M & Kimmage P, 2011. The ventilator stopped... *The Sunday Times*.

Harries R, 1995. Evidence for God's Love. In: Harries R (ed), *Questioning Belief*. London: SPCK.

Harris J, 1983. In vitro fertilisation: The ethical issues. *Philosoph*

Quart 33: 225.

Hartman L, 1997. *Into The Name Of The Lord Jesus.* Edinburgh: Clark.

Hawkes K, O'Connell J, Jones N et al, 1998. Grandmothering, menopause, and the evolution of human life histories. *Proc Nat Acad Sciences [USA]* 95: 1336-1339.

Hebblethwaite B, 1996. *The Essence of Christianity.* London: SPCK.

Heidegger, 1959. *Introduction to Metaphysics.*

Henneberg M, Eckhardt R, Chavanes S, Hssu K, 2014. Evolved developmental homeostasis disturbed in LB1 from Flores, Indonesia, denotes Down syndrome and not diagnostic traits of the invalid species Homo *floresiensis. Proc Nat Acad Sciences [USA]* 111: 11967-11972.

Hennig, H, 2014. Synchronization in human musical rhythms and mutually interacting complex systems. *Proc Nat Acad Sciences [USA]* 111: 12974-12979.

Henrikson J-O, 2011. The Erotic Self And The Image Of God. In: Huyssteen W & Wiebe E (eds), *In Search Of Self: Interdisciplinary Perspectives On Personhood.* Grand Rapids: Eerdmans 261.

Hepper P, Shahidullah S, 1994. The beginnings of mind – evidence from the behaviour of the fetus. *J Reprod Inf Psychol* 12: 143-154.

Hewson B, 2002. SPUC and the morning-after pill saga. *New Law J* 152: 1004.

Heyes C, 1993. Anecdotes, training, trapping and triangulating: do animals attribute mental states? *Animal Behav* 46: 177-188.

Heyes C, 1994. Reflections on self-recognition in primates. *Animal Behav* 47: 909-919.

Hooker M, 1997. *The Signs of a Prophet.* London: SCM Press.

Hughes G, SJ, 2007. *Is God to Blame?* Dublin: Veritas Press.

Hughes GW, SJ, 1985. *The God Of Surprises.* London: Darton, Longman & Todd.

Hull J, 1990. *Touching the Rock: An Experience of Blindness.*

London: SPCK.

Human Development Anatomy Center (National Museum of Health & Medicine, Washington, DC). http://www.lifeissues.net/writers/irv/irv_123carnegiestages1. html; http://nmhm.washingtondc.museum/collections/hdac/stage1. pdf]

Hursthouse R, 1987. *Beginning Lives*. Oxford: Blackwell.

Huxley A, 1928. *Point-Counter-Point*. London: Penguin.

Huxley J, 1941. *'The Uniqueness of Man'*. London: Scientific Book Club.

Jardine C, *The Daily Telegraph* 2011. 'Fiona Pilkington: will we hear the next cry for help?'

Johnson A, 1961. *The One and the Many in the Israelite Conception of God*. Cardiff: Cardiff University Press.

Johnson A, 1964. *The Vitality of the Individual in the Thought of Ancient Israel*. Cardiff: Cardiff University Press.

Jones P, Baylin S, 2007. The epigenomics of cancer. *Cell* 128: 683-692.

Jungel E, 1975. *Death – The Riddle and the Mystery*. Edinburgh: St Andrew's Press.

Keller C, 2011. Enigmatic Experiences: Spirit, Complexity, and Person. In: van Huyssteen W & Wiebe E (eds), *In Search Of Self: Interdisciplinary Perspectives On Personhood*. Grand Rapids: Eerdmans.

Kelsey D, 2009. *Eccentric Existence: A Theological Anthropology*. Louisville: John Knox Press 275; 284ff; 1045-1051.

Kendrick K, da Costa A, Leigh A, Hinton M, Peirce J, 2001. Sheep don't forget a face. *Nature* 414: 165-166.

Kennett J, 2002. Autism, empathy and moral agency. *The Philosoph Quart* 52: 340-352.

Kieffer J, Colgan P, 1992. The role of learning in fish behaviour. *Rev Fish Biol & Fisheries* 2: 125-143.

Kissane D & Kelly B, 2000. Demoralisation, depression and desire for death: problems with the Dutch guidelines for euthanasia of the mentally ill. *Austr NZ J Psychiatr* 34: 325-33.

Kliff S, 2011. *The Year of the Abortion Restrictions.* The Washington Post (29 December) Neylan TC, 1999. Frontal Lobe Function: Mr. Phineas Gage's Famous Injury. *J Neuropsych Clin Neuroscience.* 11: 280-283.

Krause J, Lalueza-Fox C, Orlando L, et al, The derived FOXP2 variant of modern humans is shared with Neanderthals. *Curr Biol* 17: 1908-1912.

Kuhl P, 2004. Early language acquisition: cracking the speech code. *Nat Rev Neurosci* 5: 831-843.

Kuzawa C, Chugani H, Grossman L, et al, 2014. Metabolic costs and evolutionary implications of human brain development. *Proc Nat Acad Sciences [USA]* 111: 13010-13015.

Ladd, GE, 1975. *A Theology of the New Testament.* Michigan: Eerdmans 457.

Lalueza-Fox C, Rosas A, Estalrrich A, et al, 2010. Genetic evidence for patrilocal mating behaviour among Neanderthal groups. *Proc Nat Acad Sciences [USA]* 108: 250-253.

Lanctot C, Cheutin T, Cremer M, Cavalli G, Cremer T, 2007. Dynamic genome architecture in the nuclear space: regulation of gene expression in three dimensions. *Nature Rev Genetics* 8: 104-115.

Lathif N, Phipps E, Alton P, Sharma D, 2014. Prevalence of psychiatric disorders following brain injury. *Br J Med Pract* 7: a722.

Laureys S, Faymonville M, Luxen A, et al, 2000. Restoration of thalamocortical connectivity after recovery from persistent vegetative state. *Lancet* 355: 1790-1791.

Lewis CS, 1977. *Surprised by Joy.* London: Fount Books.

Liberman A & Whalen D, 2000. On the relation of speech to language. *Trends Cogn Sci* 4: 187-195.

Lieberman P, 1969. Vocal tract limitations on the vowel repertoires of rhesus monkeys and other nonhuman primates. *Science* 164: 1185-1187.

Locke J, 1995. *An Essay Concerning Human Understanding*. New York: Prometheus Books 246-250.

Lockwood M, 1985. *Moral Dilemmas in Modern Medicine*. Oxford: Oxford University Press.

London: Demos 2012. *The Current Legal Status of Assisted Dying is Inadequate & Incoherent*.

Lorthongpanich C, Doris T, Limviphuvadh V, et al, 2012. Developmental fate and lineage commitment of singled mouse blastomeres. *Development* 139: 3722-3731.

Lucas JR, 1979. Wilberforce and Huxley: A legendary encounter. *The Historical Journal* 22: 313-330.

MacQuarrie J, 1983. *In Search of Humanity: A Theological & Philosophical Approach*. New York: Crossroad.

Magee JG & Garnett S, 1989. High Flight. In: *The Complete Works Of John Magee, Pilot Poet*. Cheltenham (UK): This England Books.

Magnusson S, 2014. *Where Memories Go: Why Dementia Changes Everything*. London: Two Roads (Harcourt Brace).

Mampe B, Friederici A, Christophe A, Wermke K, 2009. Newborns' cry melody is shaped by their native language. *Curr Biol* 19:1994-1997.

Marler P, 1991. Song learning behaviour: the interface with neuroethology. *Trends Neurosci* 14: 199-206.

Marsh MN, 2010. *Out-of-Body & Near-Death Experiences: Brain-State Phenomena or Glimpses of Immortality?* Oxford: Oxford University Press 237.

Marsh MN, 2012. *The Moral Status of the Human Embryo-Foetus: Biomedical Perspectives*. (Ethics #166). Cambridge: Grove Publications 15-17.

Marsh MN, 2013. Empathy: Mirroring Another's Predicament – Or More? *Antonianum* 3: 409-430.

Marsh MN, 2014. The Debate On Assisted Dying (AD). The Church Times, 28th March 16.

Marsh MN, 2015. Hey! What's that Gorilla doing over there? The

Illusory and Hallucinatory Nature of Day-to-Day Living. *European Review,* in press.

Marsh MN, 2015. Te Lucis Ante Terminum: A Perspective on Assisting Suicides. In: Badham P (ed), Assisted Dying: For and Against. *Modern Believing* 56: 181-194.

Martin-Achard R, 1960. *From Death to Life. A Study of the Development of the Doctrine of the Resurrection in the Old Testament.* London: Oliver & Boyd [ET: J Penney-Smith].

Martinez I, Quam R, Rosa M, et al, 2008. Auditory capacities of human fossils: a new approach to the origin of speech. *J Acoust Soc Am* 123: 3606 (doi:10.1121/1.2934784).

McMahon J, 2002. *The Ethics of Killing.* Oxford: Oxford University Press.

Melcher S, 1988. Visualising the perfect cult: the priestly rationale for exclusion. In: Eiesland N & Saliers D (eds), *Human Disability and the Service of God.* (Nashville: Abingdon Press.

Mellars P, French J, 2011. Tenfold population increase in western Europe at the Neanderthal-to-Modern human transition. *Science* 333: 623-627.

Meyer M, Kircher M, Gansuage M, et al, 2012. A high-coverage genome sequence from an archaic denisovan individual. *Science* 338: 222-226.

Midgley M, 1983. Selfish genes and social Darwinism. *Philosophy* 365-377.

Midgley M, 1992. *Science as Salvation.* London: Routledge.

Miranda J, 1977. *Being & The Messiah* [ET]. Maryknoll: Orbis.

Misteli T, 2007. Beyond the sequence: cellular organization of genome function. *Cell* 128: 787-800.

Misteli T, 2011. The inner life of the genome. *Scientific American* 304: 48-53.

Mithen S, 2005. *The Singing Neanderthals: The Origins of Music, Language, Mind and Body.* London: Weidenfeld & Nicolson.

Muchembled R, 2011. *A History of Violence: From the End of the Middle Ages to the Present.* Cambridge: Polity Press.

Muddiman J, 1994. 'I believe in the Resurrection of the Body'. In: Barton S & Stanton G (eds), *Resurrection*. London: SPCK 135.

Muller-Hill B, 1988. *Murderous Science: Elimination by Scientific Selection of Jews, Gypsies, and Others, Germany 1933-1945*. Oxford: Oxford University Press.

Murphy N, 2002. The Resurrection Body And Personal Identity: Possibilities And Limits Of Eschatological Knowledge. In: Peters T, Russell RJ, Welker M, *Resurrection*.

Newton M, 2002. *Savage Girls and Wild Boys*. London: Faber & Faber.

Northcutt RG, 1996. The Agnathan Ark: the origin of craniates brains. *Brain Behav Evol* 48: 237-247.

O'Bleness M, Searles V, Varki A, Gagneux P, Sikela J, 2012. Evolution of genetic and genomic features unique to the human lineage. *Nature Rev Genet* 13: 853-866.

O'Donovan O, 1984. *Begotten or Made?* Oxford: Clarendon Press.

O'Regan JK, Noë A, 2001. A sensorimotor account of vision and visual consciousness. *Behav Brain Sci* 24: 9390-1031.

Orbach S, 2009. *Bodies*. London: Profile Books.

Pannenberg W, 1971. *Basic Questions in Theology* (Vol 3). London: SCM Press.

Pannenberg W, 1972 (1962) *What is Man?* Philadelphia: Fortress Press.

Pannenberg W, 1996. Baptism as remembered 'Ecstatic' Identity. In: Brown D & Loades A (eds), *Christ: The Sacramental Word*. London: SPCK.

Paul K, 2015. The ARS MORIENDI: A Practical Approach to Dying. *Modern Believing* 56: 207-220.

Peters J, 2014. The role of genomic imprinting in biology and disease: an expanding view. *Nature Rev Genet* 15: 517-530.

Peters T, Russell RJ, Welker M, 2002. *Resurrection: Theological And Scientific Assessments*. Grand Rapids: Eerdmanns.

Pinker S, 2011. *The Better Angels of our Nature: The Decline of Violence in History and its Causes*. London: Allen Lane.

Pistorius M, 2011. *Ghost Boy.* London: Simon & Schuster.

Pitcher G, 2010. *A Time to Live: The Case against Euthanasia & Assisted Dying.* Oxford: Monarch Books.

Portela A, Esteller M, 2010. Epigenetic modifications and human disease. *Nature Biotechnology* 28: 1057-1068.

Prigogone I, 1996. *The End of Certainty: Time, Chaos, and the New Laws of Nature.* New York: The Free Press 17.

Quam R & Rak Y, 2008. Auditory ossicles from south-western Asian Mousterian sites.

Raine A, Buchsbaum M, LaCasse L, 1997. Brain abnormalities in murderers indicated by positron emission tomography. *Biol Psychiatr* 42: 495-508.

RexvBourne [1939] 1 K.B. 687.

Richards M, Trinkaus E, 2009. Isotopic evidence for the diets of European Neanderthals and early modern humans. *Proc Natl Acad Sciences [USA]* 106: 16034-16039.

Richardson B, 2007. Primer: epigenetics of autoimmunity. *Nature Clin Pract Rheumatol* 3: 521-527.

Rifkin J, 2009. *The Empathic Civilisation: The Race towards a Global Consciousness.* Cambridge: Polity Press.

Robinson J, 1957. *The Body: A Study in Pauline Theology.* London: SCM Press 1957.

Rodriguz-Vidal J, d'Errico F, pacheo G, et al, 2014. A rock engraving made by Neanderthals in Gibraltar. *Proc Nat Acad Sciences [USA]* 111: 13301-13306.

Rolnick P, 2007. *Person, Grace, And God.* Grand Rapids: Eerdmans.

Sacks O, 1995. *An Anthropologist on Mars.* London: Picador.

Scruton R, 1996. Sex. In: *An Intelligent Person's Guide to Philosophy.* London: Duckworth 128-129.

Scruton R, 2011. Neurononsense And The Soul. In: van Huyssteen J & Weibe EP (eds), *In Search Of Self.* Grand Rapids: Eerdmans 338-356.

Seth AK, Baars BJ, Edelman DB, 2005. Criteria for consciousness in humans and other mammals. *Consc Cogn* 14: 119-139.

Singer P, 1979. *Practical Ethics.* Cambridge: Cambridge University Press.

Singleton J, 1994. Ethical principles at the beginning of life. *J Reprod Inf Psychol,* 12: 139.

Smith Wesley J, 2000. Culture *of Death: The Assault on Medical Ethics in America.* New York: Encounter Books.

Stacey D, 1956. *The Pauline View of Man, in Relation to its Judaic and Hellenistic Background.* London: Macmillan.

Stevens K, House A, 1995. Development of a quantitative description of vowel articulation. *J Acoust Soc Am* 27: 484-493.

Studdert Kennedy G, 1986 (1929). Indifference. In: *The Unutterable Beauty.* Oxford: Mowbray.

Tallis R, 1999. *The Explicit Animal: A Defence of Human Consciousness.* Basingstoke: Macmillan Press.

Tallis R, 2011. *Aping Mankind: Neuromania, Darwinitis and the Misrepresentation of Humanity.* Durham: Acumen.

Tate A, Fischer H, Leigh A, Kendrick K, 2006. Behavioural and neurophysiological evidence for face identity and face emotion processing in animals. *Phil Trans R Soc B* 361: 2155-2172;

Thomson JJ, 1971. A defense of abortion. *Philosophy & Public Affairs* 1: 47-66.

Tillich P, 1997. *Systematic Theology* (Vol I). London: SCM Xprint 164.

Tooley M, 1983.*Abortion and Infanticide.* Oxford: Oxford University Press.

Tran L, Hino H, Quach H, et al, 2012. Dynamic microtubules at the vegetal cortex predict the embryonal axis in zebrafish. *Development* 139: 3644-3652.

Tversky A and Kahneman D, 1974. Judgement under Uncertainty: Heuristics and Biases. *Science* 185: 1124-1131.

Urbano F, D'Onofrio S, Luster B, et al, 2014. Pedunculopontine nucleus gamma band activity – preconscious awareness, waking, and REM sleep. *Front Neurol* 5:1-12.

van Huyssteen JW, 2006. *Alone in the world? Human Uniqueness in*

Science and Theology. Grand Rapids: Eerdmans.

van Huyssteen W & Wiebe E, 2011. *In Search Of Self: Interdisciplinary Perspectives On Personhood.* Grand Rapids: Eerdmans.

Vanhaudenhuyse A, Noirhomme Q, Tshibanda L, et al, 2010. Default network connectivity reflects the level of consciousness in non-communicative brain-damaged patients. *Brain* 133: 161-171.

Vanstone WH, 1982. *The Stature of Waiting.* London: Darton, Longman &Todd.

Vassena R, Boue S, Gonzalez-Roca E, et al, 2011. Waves of early transcriptional activation and pluripotency program initiation during human preimplantation, *Development* 138: 3699-3709.

Ward L & Butt R, 2007. *The Guardian,* 24 October.

Wark P, 2009. *The Times* (London), 24 Sept. 7.

Warnock GJ, 1971. *The Object of Morality.* London: Methuen.

Warnock M, 1984. Report of the Committee of Enquiry into Human Fertilisation and Embryology. London: Her Majesty's Stationary Office, 1984.

Warren MA, 1973. On the moral and legal status of abortion. *The Monist* 57: 43-61.

Watson J, Crick F, 1953. Molecular structure of nucleic acids: a structure for deoxynucleic acid. *Nature* 171: 737-738.

Weaver T, Hublin J-J, 2009. Neanderthal birth canal shape and the evolution of human childbirth. *Proc Nat Acad Sciences [USA]* 106: 8151-8156.

Webster J, 2005. *Confessing God: Essays in Christian Dogmatics II.* London: Clark International.

Whitaker M, 2005. Syngamy and cell cycle control. In: Myers RA (ed), *Encyclopaedia of Molecular Cell Biology and Molecular Medicine.* Weinham: Wiley Verlag.

White S, Fisher S, Geschwind D, et al, 2008. Singing mice, songbirds, and more: models for FOXP2 function and dysfunction in human speech and language. *J Neurosci* 26: 10376-10379.

Wiessner P, 2014. Embers of society: firelight talk among the Ju/'hoansi Bushmen.

www.pnas.org/cgi/doi/10.1073/pnas.1404212111.

Wilkes K, 1984. Is consciousness important? *Brit J Philosoph Sci* 35: 223-243.

Williams Glanville, 1958. *The Sanctity of Life & the Criminal Law*. London: Faber & Faber.

Wittgenstein L, 1918. *Tractatus Logico-Philosophicus*. USA (Pbk).

World Health Organisation, at

http://who/int/topics/disabilities/en/

Wright NT, 2003. *The Resurrection of the Son of God*. London: SPCK.

Young F, 1984. *Can These Dry Bones Live?* London: SCM Press 57-59.

Young F, 1985. *Face to Face*. London: Epworth Press.

Zhang Y, Lanback P, Vibranovski M, Long M, 2011. Accelerated recruitment of new brain development genes into the human genome. *PLoS Biol* e1001179.

Zizioulas J, 1975. Human capacity and incapacity. *Scottish Journal Theology* 28: 401-447.

Zizioulas J, 1985. *Being as Communion*. New York: S Vladimir's Seminary Press.

Zizioulas J, 1991. On Being a Person: Towards an Ontology of Personhood. In: Schwöbel C and Gunton C (eds), *Persons, Divine and Human*. Edinburgh: Clark.

Zizioulas J, 2006. *Communion And Otherness: Further Studies In Personhood And The Church,* (ed: McPartlan P). London: Clark.

Zizioulas J, 2011. *The Eucharistic Communion And The World* (ed. Tallon LB). London: Clark.

Glossary

Alterity: Otherness, or a state of being other or different.

Blastocyst: The cavitated collection of blastomeres which characterises the embryo before starting to embed into the uterine wall.

Blastomere: The initial cells produced after the zygote begins to divide. These are multi-potential cells, until further development channels them into specific, differentiated cell types. There are at least 250 differentiated types of cell within the human body.

Broca: From careful post-mortem brain examinations, Paul Broca (1824-1888) discovered an area in the lower, rear aspect of the left frontal lobe which seemed important in articulating speech. Known as Broca's area.

Chalcedon: The definition of the church (AD 451) about Jesus as truly man and truly God, two natures co-existing without confusion, change, division, separation, but combined into one person (prosopon) or entity (hupostasis).

Deontology/Deontological: An ethical system based on obligations, responsibilities and duties, like The Commandments or any secular legal system of rules and laws.

Diachronic: The historical or evolutionary change of something.

Diaspora: A dispersion of any people or race from its origins.

Discordance: Monozygotic twins, although deriving genetically from the same ovum and sperm, often differ in post-natal life. Twins are used to identify genetic risks for certain diseases through observations on those who develop a certain disease, and those who do not. This gap in disease susceptibility among twins is termed discordance, and can be quantified.

Entropy: The degree of disorder in a system. The Second Law of Thermodynamics states that the entropy in any system must only increase.

Epigenetics: The concept of non-Darwinian influences on the genes through, for example, environmental causes or influences.

Eschaton and Eschatology: These are theological terms denoting, and concerned with, the 'end times' when Creation as we know it will be wound up, with the Second Coming of the Messiah in glory, from the Greek *eschaton*. This is to be distinguished from an end, aim or purpose, as defined by teleology, from the Greek *telos*.

Exousia: Loosely, the power of God and the exhibition of its authority and freedom of action, as exemplified by Jesus throughout his ministry on earth. In Greek, 'ousia' means the divine nature, so ex (out of) suggests something coming or derived from the Godhead itself.

Genotype: A term used to refer to all the chromosomes inherited by any individual and the genes that are carried by them.

Godhead: In Christian theology, the Godhead comprises God-the-Father, God-the-Son, and God-the-Holy-Spirit, or the Trinity. Each of these three within the Trinity is known as an Hypostasis or a Person. In the Orthodox tradition, it is believed that baptism admits the biologically fashioned human into the Godhead, and is thereby 'en-hypostatised' by that action, through the Spirit and in Christõ.

Ipseity: One's selfness or unique being.

Metanoia: From the Greek metanoe, meaning to turn around, change one's mind, and metaphorically, to repent.

Morula: Describes the roughly spheroidal collection of blastomeres, before it cavitates.

Mutation: An alteration in one or more bases of an individual's DNA. This results in a changed product of that gene mutation which may either disrupt normal physiology, or enhance it. For example, the FOXP2 gene undergoes two mutations which seem to influence the articulatory aspects of speech in humans.

Ontogeny: The process of development, anatomical and/or behavioural, of a specific or individual organism from its origin to maturity.

Ontology: Ontos is the participle from the Greek verb 'to be', meaning being something or having an existence. It can refer to persons and also to abstract qualities such as faith, hope, or history. It is a branch of metaphysics dealing with the nature of being.

Ossicles: The small bones in the middle ear which transmit vibrations from the tympanic membrane to the auditory division of the 8th cranial nerve, thereby subserving hearing and the interpretation of the sounds impinging onto the outer ear.

Phenotype: The body and other characteristics which constitute an individual, resulting from the interplay between genes and environment.

Phylogeny: Refers to the evolutionary history of a group or species of organisms.

Post-exilic: Refers to the time when the Jews were restored to Israel by Cyrus, with the re-establishment of the Temple (539 BCE).

Pre-exilic: The period before the exile of the Israelites to Babylon (597 BCE).

Primitive streak: At around 14 days post-fertilisation, the human blastocyst develops a linear furrow along the presumptive front-to-back axis of the developing embryo. The appearance is produced by an infolding of the outer coat of cells (termed epiblasts) which give rise to skin cells and those cells lining the inner mucosal surfaces of the body. The streak allows migration of additional cells along this furrow, in order to give rise to the spinal cord.

Septuagint: The translation of the Hebrew Old Testament into Greek (~200 BCE). Said to have been translated by 70 (or 72) Jewish scholars, hence its other designation as 'LXX'.

Synchronic: Pertaining to the present time.

Transcendence: The concept of being able to extend, or be extended, beyond oneself, to reach out or to cross barriers, thus even to encounter the divine.

Utilitarianism: The ethical system that states that actions are right if they secure the greatest happiness for the largest number of people. See Deontology.

Vestibular: A complex brain system which begins in the inner ear (cochlea, saccule and utricle) and which transmits signals through the vestibular division of the 8th cranial nerve to higher centres in the brain. This allows subjects to be orientated in space and towards gravitational forces, and to know where they are with respect to the environment.

Wernicke: It was Carl Wernicke (1848-1904) who discovered a part of the brain in the left temporal lobe concerned with inability to recognise or interpret heard speech (so-called 'sensory aphasia'), and famously termed Wernicke's speech area.

Zygote: The cell immediately resulting from fusion of sperm with ovum. Further divisions give rise to blastomeres.

Books by the Same Author

The Immunopathology of the Small Intestine. Chichester: Wiley 1987

Coeliac Disease. Oxford: Blackwell Scientific Publications 1992

Coeliac Disease. (Monographs in Molecular Medicine, #41). Totowa [NJ]: Humana Press 2002

Out-of-Body & Near-Death Experiences: Brain-State Phenomena or Glimpses of Immortality? Oxford: Oxford University Press 2010

The Moral Status of the Human Embryo-Foetus: Biomedical Perspectives. Cambridge: Grove Publications 2012

Index of Names

Index of Subjects

BOOKS

Iff Books is interested in ideas and reasoning. It publishes
material on science, philosophy and law. Iff Books aims to work
with authors and titles that augment our understanding of the
human condition, society and civilisation, and the world or
universe in which we live.